Deepwater Petroleum Exploration & Production

A Nontechnical Guide

Deepwater Petroleum Exploration & Production

A Nontechnical Guide

William L. Leffler
Richard Pattarozzi
Gordon Sterling

Copyright © 2003 by
PennWell Corporation
1421 South Sheridan
Tulsa, Oklahoma 74112 USA

800.752.9764
1.918.831.9421
sales@pennwell.com
www.pennwell-store.com
www.pennwell.com

Managing Editor: Marla Patterson
Production Editor: Sue Rhodes Dodd
Cover design: Shanon Moore

Library of Congress Cataloging-in-Publication Data Pending
ISBN 0-87814-846-9

Printed in the United States of America

1 2 3 4 5   07 06 05 04 03

Contents

List of Illustrations vii

Foreword xi

Introduction xiii

List of Acronyms xv

Chapter 1 – A Century Getting Ready 1
  Oily Beginnings 1
  Free At Last 4
  Other Humble Beginnings 5
  Superior Approaches 6
  Divers and ROVs 16
  Geology, Geophysics, and Other Obscure Sciences 18
  Permanence 23
  The Learning Curve Bends Over 25

Chapter 2 – Letting Go of the Past 27
  The Confusion of the 1980s 27

Chapter 3 – Exploring the Deepwater 35
  Identifying the Prospect 39
  Drilling a Wildcat 47
  Deepwater Plays in Context 48
  Geology—the Shelf vs. the Deepwater .49

Chapter 4 – Drilling and Completing Wells 53
  The Well Plan 54
  Rig Selection 55
  Drilling 57
  Completing the Well 61
  Special Problems 66

Chapter 5 – Development Systems 69
  Development System Choices 70
  Choosing Development Systems 72

Chapter 6 – Fixed Structures 77
  The Conventional Platform 77
  The Concrete Platform 78
  The Compliant Tower 79
  From Here to There 81

Installing Platforms 83
Installing Concrete Gravity Platforms 85
Setting the Deck 85
Setting the Pipeline Riser 87

Chapter 7 – Floaring Production Systems 89
Tension Leg Platforms (TLP) 91
Monocolumn TLP 92
FPSO 93
FDPSO 96
FSO 96
FPS 97
Spars 98
Construction and Installation 100
Mooring Spreads 104

Chapter 8 – Subsea Systems 107
Wells 109
Trees 109
Manifolds and Sleds 110
Flowlines, Jumpers, and Gathering Lines 1111
Umbilicals and Flying Leads 112
Control Systems 113
Flow Assurance 114
System Architecture and Installation 115
ROVs 119

Chapter 9 – Topsides 125
Oil Treatment 127
Water Treatment 128
Gas Treatment 129
Personnel and Their Quarters 133
Safety Systems 134
Auxiliary Systems 136

Chapter 10 – Pipelines, Flowlines, and Risers 139
The Boon and Bane of Buoyancy 141
Laying Pipe 142
Bottom Conditions 147
Risers 148
Pipeline System Operations 153

Chapter 11 – Technology and the Third Wave 155

Index 153

List of Illustrations

F–1   Layout of Deepwater.............................. xi

1–1   Piers and derricks at Summerland, California, 1901 2

1–2   Drilling from wooden pile platforms in Lake Caddo, Texas 3

1–3   The submersible *Giliasso* 4

1–4   Superior's prefabricated template platform 6

1–5   Kerr–McGee's platform in the Ship Shoal area of
the Gulf of Mexico 7

1–6   Submersible rigs.................................... 8

1–7   Early jack-up rigs.................................... 9

1–8   Aboard the *Scorpion,* first jack-up to use rack
and pinion drives.................................... 10

1–9   The drilling sequence used aboard the *CUSS 1*.............. 12

1–10   Pioneer Bruce Colipp's DaVinciesque Diagram 13

1–11   The drillship *Eureka*................................ 14

1–12   The *Mobot* 17

1–13   Twin cranes lifting a jacket into place.................... 19

1–14   Early offshore seismic collection 20

1–15   Fixed platforms by installation year 22

1–16   The *Bullwinkle* platform being towed to sea 23

1–17   The *Bullwinkle* platform in place 24

1–18   Exploration and Production—the first and second waves 25

2–1   Rig Count in the Gulf of Mexico, 1959–82 27

2–2   Average size of fields discovered in the Gulf of Mexico 29

2–3   Oil and gas production in the Gulf of Mexico 29

2–4   Petrobras discoveries and drilling records 31

2–5   Deepwater—the third wave 34

3–1   The exploration part of the exploration and
production process................................. 35

3–2   Reservoir rock, trap, sealing mechanism, and the
migration of hydrocarbons.......................... 36

3–3   Geologists and geophysicists at a work station 40

3–4   The seismic vessel *Seisquest* 41

3–5   Offshore seismic acquisition. 42

3–6   2D Seismic display of a hanging wall anticline 42

3–7   Migration—reflections from updip location 44

3–8   3D block of seismic data with horizontal and vertical slices. 45

3–9   Full color display of an oil and gas field 46

3–10   Seismic display in a visualization room 46

3–11   Profile of the Gulf of Mexico 50

3–12   Salt-related formations: an anticline, a dome, and faulting 51

4–1   The drilling process 53

4–2   Cover sheet for a drilling prognosis 54

4–3   The semisubmersible *Nautilus* 56

4–4   The drillship *Enterprise* 56

4–5   Blowout preventer system 60

4–6   Well completion process 62

4–7   Cover sheet for a well completion prognosis. 62

4–8   Downhole gravel pack 65

4–9   Loop current and eddy currents in the Gulf of Mexico 67

5–1   Steps in development of deepwater projects 69

5–2   Development system options for deepwater projects 70

5–3   The *Mensa* field, a subsea development 74

6–1   Fixed steel platform 78

6–2   Concrete gravity structure 78

6–3   Compliant tower 79

6–4   Construction cranes "rolling up" a steel frame
       for a deepwater jacket 80

6–5   Concrete gravity structure being towed to an installation site . . . 82

6–6   Steel jacket launching from a barge 83

6–7   Steel jacket connected to a piling with grout. 84

6–8   *Saipem 7000* heavy lift barge setting a deck section 86

6–9 Topsides deck ready to be lowered onto legs of a concrete structure 87

7–1 Floating system options for deepwater projects 89

7–2 Tension leg platform *Auger*, sometimes called "The TLP that ate New Orleans" 90

7–3 TLP for the *Brutus* field in the Gulf of Mexico 91

7–4 TotalFinalElf *Matterhorn SeaStar©*, a monocolumn TLP 92

7–5 The FPSO *Anasuria* in the North Sea 93

7–6 An FPSO internal turret 94

7–7 An FPSO cantilevered turret 94

7–8 The *Na Kika* development scheme uses an FPS 97

7–9 Three spar options—conventional, truss, and cell........... 98

7–10 The *Genesis* Spar en route to the wellsite 99

7–11 TLP hull under construction 101

7–12 Hull of TLP *Brutus* en route to installation on a heavy lift vessel................................. 102

7–13 The *Genesis* Spar uprighting from its floating position 103

7–14 A 500-lb chain link for a mooring chain 105

7–15 Mooring spread for the *Na Kika* FPS 106

8–1 Subsea development scheme for *Angus*, hooked up to *Bullwinkle* 108

8–2 The *Na Kika* development scheme for six subsea fields 108

8–3 A subsea tree with ROV-friendly connections 110

8–4 The *Mensa* subsea field manifold, jumpers, flying leads, and sled 111

8–5 Cutaway view of an electro-hydraulic umbilical 112

8–6 A subsea tree component being lowered from the back of a work boat.................................... 116

8–7 Several vessels working simultaneously at the Crosby project .. 116

8–8 A crane vessel lowering a subsea manifold 117

8–9 Work class ROV 120

8–10 ROV in its cage being launched 121

8–11   Sequence of an ROV attachment of a pipeline to a
       subsea manifold 122

8–12   ROV training simulator 123

9–1   Crude oil processing on deepwater platforms 125

9–2   Multistage gas separation on deepwater platforms 127

9–3   Natural gas processing on deepwater platforms 130

9–4   Glycol gas dehydration column 131

9–5   Crew quarters on a deepwater platform 134

9–6   A survival capsule aboard a TLP 135

9–7   Topsides layout of a fixed platform 136

9–8   Exploded view of *Mars* TLP topsides 137

9–9   Deck module being lifted onto the *Mars* TLP 138

10–1   Well production flows from the wellhead to
       shoreline connections. 139

10–2   Flexible pipe construction 140

10–3   S-lay method for deepwater pipelines 141

10–4   S-lay vessel, *Solitaire*. 143

10–5   J-lay barge method for deepwater pipelines 144

10–6   J-lay tower on the crane vessel *Balder*. 144

10–7   Vertical and horizontal reel barge methods 145

10–8   Reel barge at a yard 145

10–9   Reel barge *Hercules* laying pipe 146

10–10 Variations on pipeline tow-in installation 147

10–11 Attached and pull tube risers 148

10–12 Definition of a catenary. 149

10–13 Top tensioned risers. 150

10–14 Riser connections at the *Auger* TLP 150

10–15 Motion compensation mechanism for top tension risers 151

10–16 Some flexible riser configurations. 152

11–1   Deepwater areas with hydrocarbon potential 157

Foreword

Forewarned forearmed.

From *Don Quixote*, Miguel De Cervantes, 1547–1616

We needed to hurry up and write this book. Our first two chapters bring you up to date from the first geologic toe in the water in California a hundred years ago to stepping off the Outer Continental Shelf of the Gulf of Mexico into thousands of feet of water as well as the plunge into the Campos Basin off the coast of Brazil. As with any forage into a new frontier, history will occur almost every day after our publication date. Still, the journey, as far as we take you, is a worthwhile prelude to understanding even future deepwater operations. (See Fig. F–1)

The process for exploring for, developing, and producing petroleum in the deepwater is no different than for the shelf or really, the onshore. From the outside, just four steps take place—explore, appraise, develop, and produce. From the inside, each of these steps takes another handful or even dozens of steps, depending on how closely we look. And we look closer in the next three chapters.

Following all this how and who, the next six chapters examine in detail the engineering and scientific schemes that companies use in the deepwater, dealing especially with how they differ from shallower operations and the onshore.

In this book, we try to take you down a funnel, the one shown in the accompanying figure, from perspective to insight to process to application. At the risk of rapid obsolescence, we have added a last chapter on future challenges as we see them.

We had to assume something about your knowledge of E&P operations—onshore and

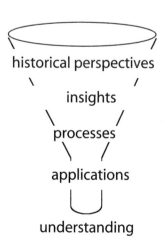

historical perspectives

insights

processes

applications

understanding

Fig. F–1 Layout of Deepwater

on the shelf. Since you bought this book which has the words Nontechnical Guide in the title, we treat each subject as if you have but a modicum of background. You should find almost everything easy to understand. If you need more depth, our publisher, PennWell, has a few other books such as Norm Hyne's Nontechnical Guide to Petroleum Geology, Exploration, Drilling, and Production to help you out.

This book is a collaborative effort of two engineers and a business guy. In addition, we had the invaluable input from a throng of industry experts and former colleagues. We can recognize a few: Howard Shatto, Bruce Collipp, Mike Forrest, Jim Day, Dick Frisbie, Ken Arnold, Doug Peart, George Rodenbusch, Susan Lorimer, Jim Seaver, Bob Helmkamp, Mitch Guinn, Gouri Venkataraman, Alex van den Berg, Don Jacobsen, Harold Bross, Martin Raymond, and Frans Kopp. Without their help, we could not have satisfied our own standards for a quality product. Still, we interpreted all they said and are therefore responsible for the way we presented it.

W. L. Leffler
R. A. Pattarozzi
G. Sterling

Introduction

My reaction when Rich Pattarozzi told me he was working on a nontechnical book about the development and production of oil and natural gas in deepwater has not changed: "At last. Without doubt, our industry needs this book. I could not be more enthusiastic about its content."

Going into the deepwater demands so much of so many, that few individuals can grasp all of the intricate details and technical challenges that have to be overcome. I believe these three authors are unsurpassed in their ability to tell the story from start to finish in an understandable fashion.

Rich Pattarozzi, the talented senior executive at Shell Exploration and Production who created our deepwater organization, provided the dynamic leadership to take Shell where no oil and gas company had been. For Rich, technical and economic challenges were never roadblocks. They were merely opportunities for creativity and innovation that brought out the best in Shell's staff. Gordon Sterling pioneered many of the technical breakthroughs required to take our company on its incredible journey. Never afraid to question conventional engineering paradigms, he encouraged and nurtured the new and often radical approaches necessary to break through the technical barriers that inevitably occurred along the way. And finally, there is Bill Leffler, a long-time planner and strategic thinker at Shell, who has a gift of communicating through the written word. Despite his nontechnical background, Bill is able to transform complicated concepts into clear and concise words that are understandable for the expert and for the lay person.

While this book is all about the oil and gas companies' operations in deepwater, no doubt it will find a home on the desks and bookshelves of many non-oil company readers. Our industry has been most fortunate to have the thousands of dedicated service and supply personnel whose help and innovation in their area of expertise have made this deepwater story possible. Some of the key sectors that have made significant contributions are fabrication and construction, marine transportation, offshore drilling, producing systems, and oil and gas pipelines. Through this incredible journey, a vital partnership between the oil and gas operators and the service and supply industry has developed along the lines so evident in the 150+ year history of the oil and gas industry.

I commend this book to you, not only for its readability, but also for the story that it tells. It is the saga of thousands of men and women, working individually

1

A Century Getting Ready

In an unchanging universe

a beginning in time is something

that has to be imposed by

some being outside the universe.

From *A Brief History of Time,* Stephen Hawking (1942–)

Oily Beginnings

Most petro-historians, from whom we have unabashedly borrowed much of this chapter, trace offshore exploration and production to Summerland, California. In 1897, at this idyllic-sounding spot just southeast of Santa Barbara, Summerland's founder, a spiritualist, and sometimes wildcatter, H. L. Williams, boldly inched into the surf. With oil seeping from the ground back for hundreds of yards from the water's edge, Williams skipped the exploration stage and immediately built three wooden piers out some 450 yards from the shoreline. Water depths reached 35 ft. (*See* Fig.1–1.) Over the next three years, he erected 20 derricks atop the piers. The power generators and other supporting equipment sat along the beachfront. Williams' crew, like most other drillers at that time, had not yet adopted rotary drilling rigs. Instead, they set a steel pipe, called a casing, from the drilling platform down through the sandy bottom. Then they used cable tools to pound their way down 455 ft to two oil sands.

Bold as it was, the effort paled in comparison to its contemporary, Spindletop, the 80,000-barrels-per-day gusher drilled onshore near Beaumont, Texas. The most prolific well at Summerland

Fig. 1–1 Piers and Derricks at Summerland, California, 1901 (Courtesy USGS)

reached only 75 barrels per day, the average well only 2. Production peaked in 1902 and declined rapidly after that. Williams abandoned Summerland, the field, and his cult a few years later and left an ugly blight of piers and oily beach behind. The piers decayed slowly until 1942 when they finally succumbed to a violent tidal wave.

Scores of other venturers copied the pier and derrick technique along the California coast over the next 10 years. At one, the Elwood field, the piers extended 1800 ft from the shore, and still reached a water depth of only 30 ft. Not until 1932 did the Indian Oil Company courageously build a stand-alone platform in the shallow Pacific Ocean waters off Rincon, California.

The term *offshore* usually conjures up visions of vast expanses of water well beyond the pounding surf. However, the next important bit of offshore history happened in a more contained locale. In the area around Lake Caddo in East Texas over the years following 1900, wildcatters searching for oil continually stumbled on pockets of associated natural gas—to the chagrin of most. Gas cost much more to transport and required large discoveries and dense populations to create a market. Only one out of three of these conditions appealed to an East Texas wildcatter. In 1907, J. B. McCann, a scout for Gulf Oil Corporation, mulled over maps of the Lake Caddo area and thought about the gassy province that lay below. Late one night, he used a novel tool to prove his theories. He rowed across the lake, carefully touching lighted matches to the vapors bubbling from the waters. Besides successfully avoiding self-immolation, he convinced himself—and eventually W. L. Mellon in Gulf headquarters at Pittsburg—that a large oil and gas field crossed under the lake.

Gulf acquired the concession to drill 8000 acres of lake bottom and brought new techniques to the area and to the industry. Starting in 1910, they towed up the Mississippi and Red rivers a floating pile driver, a fleet of supply boats, and barges of derricks, boilers and generators. In the lake, they drove pilings using the abundant cypress trees felled along the shoreline. Atop they built platforms for their derricks (Fig. 1–2) and pipe racks. Each drilling/production platform had its own derrick and gas-driven generator. Each pumped production down a 3-in. diameter steel flowline laid along the lake bottom to separation and gathering stations atop other platforms.

Fig. 1–2 Drilling from Wooden Pile Platforms in Lake Caddo, Texas (Louisiana Collection, State Library of Louisiana)

Over the next four decades, Gulf drilled 278 wells and produced 13 million barrels of oil from under Lake Caddo, creating in the process a commercially successful prototype for water-based operations, the platform on piles.

Concrete progress

American notions aside, not all progress and innovation took place in the United States. Production in Lake Maracaibo, Venezuela, in the mid-1920s might have replicated Caddo Lake but for one thing, the dreaded *teredo*. These intrusive shipworms had pestered mariners since ancient times. In less than eight months, these pesky parasites could chew through the wooden pilings that supported a Lake Maracaibo drilling platform, not allowing enough time to make a profit. Creosoted pine from the United States proved a technically effective antidote but the expense made it an uneconomic solution.

In an instance of serendipity, the Venezuelan government had contracted with Raymond Concrete Pile Company to build a seawall on the lakefront near the oilfields, thereby underwriting an entire infrastructure necessary to make concrete platform pilings. Lago Petroleum (later Creole Petroleum, and then Esso, until the Venezuelan government nationalized it) tried using these concrete pilings in place of the wooden pilings. Soon they were fitting the pilings with steel heads to allow faster installation and tying them together with steel and wire rope for structural integrity. In the next 30 years, industry erected 900 concrete platforms in Lake Maracaibo. By the 1950s, they used hollow cylindrical concrete piles with 5-in. walls and 54-in. diameters, 200 ft long, and pre-stressed with steel cable.

Free at Last

At the same time Lago was developing Lake Maracaibo, the Texas Company (later Texaco) was searching for a better idea for their properties in the Louisiana swamps. Platforms on driven wooden pilings worked, but the expense left room for improvement. The idea of using a barge sunk in place as a drilling platform intrigued the Texas Company. In their own prudent way, they first visited the U.S. Patent Office and discovered that Louis Giliasso, a merchant marine captain who had worked the Lake Maracaibo fields, had already claimed the idea. After a Byzantine search, they found him in 1933, improbably running a saloon in Panama.

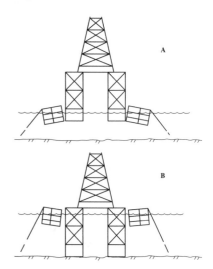

Soon after, the Texas Company sank two standard barges, side-by-side, in a swampy section of Lake Pelto, Louisiana. With only a few feet of water to deal with, they had enough freeboard to weld a platform on top and install a derrick. (*See* Fig. 1–3.) In a magnanimous moment, they named the first submersible the *Giliasso* after its inventor. They sank another barge nearby with a boiler for power supply and proceeded to drill a well to 5700 ft. Like most of their competitors by the 1930s, they used a rotary drilling rig. Undaunted by finding no hydrocarbons

Fig. 1–3 The Submersible Giliasso (from the original U.S. patent application: a – afloat; b – submerged)

and having to abandon the well, they pulled casing, refloated the barges, and quickly moved around the lake, drilling another five wells over a year's time. A triumph in innovation and efficiency, the *Giliasso* reduced lost time from completion of one well to drilling the next well from 17 days to 2. Mobile offshore drilling had begun.

Other Humble Beginnings

In the 1930s, the Pure Oil Company conducted onshore geophysical and seismic research near the coastal town of Creole, Louisiana. They concluded that the oil sands extended offshore. In 1937, they partnered with Superior Oil Company to test a 33,000-acre offshore concession they had acquired from the state of Louisiana. Brown & Root built for them an unprecedented 30,000-square-ft platform atop timber pilings in 14 ft of water a record one mile from the beach. The platform stood 15 ft above the water. With vivid memories of the hurricane that had killed 6000 people on Galveston Island just a few decades earlier, they reinforced the structure by sheer brute force using steel strapping and redundant piling. In an environment totally unprepared for a new offshore industry, the operator resorted to shrimp boats to tow the equipment barges to the site, to haul crews at the end of each shift, and to be supply boats.

The first well, drilled to 9400 ft, proved successful. Pure soon expanded the modest platform, drilled 10 more wells directionally, and eventually pulled nearly four million barrels from the Creole Field.

Shortly after this pioneering step, the oil parade picked up momentum. Humble Oil tried a similar but unsuccessful operation off McFadden Beach on the upper Texas coast in 1938. However, Humble was still unwilling to abandon the onshore paradigm and built a trestle several thousand feet out from shore, inexplicably stopping almost 100 ft from the platform. On top they placed railroad tracks, which they used to haul equipment and supplies. In 1938, a hurricane ravaged the trestle. Unshaken, they rebuilt it, but for naught, because they ultimately found no commercial oil deposits and abandoned the whole scheme.

In 1946, Magnolia Petroleum Company vaulted to a point six miles from the Morgan City, Louisiana coast. Offshore seismic and geological surveys convinced them that oil provinces unrelated to onshore finds lay in the Gulf of Mexico. Still, they worked in only 16 ft of water. They used a conventional design for their facility except for steel pilings under the area of the platform that supported the derrick, a concession to their concern about stability during harsh weather. Alas, their effort yielded no oil either.

Superior Approaches

The next year, Superior took another leap, technically, economically, and geographically. They moved 18 miles from the Louisiana coast, still in only 20 ft of water. They judged the pile-supported platform in their Creole Field too expensive to build in the new, more remote site. Instead, they had the J. Ray McDermott Company construct a steel tubular structure in an onshore yard and barged the prefabricated units to the site. Horizontal and diagonal members linked the tubulars like huge Tinker Toys (Fig. 1–4). With these innovative steps, Superior shortened installation time, improved structural integrity, reduced costs, improved safety conditions around the installation, and, to the contractors' delight, created a new industry sector: prefabrication.

Fig. 1–4 Superior's Prefabricated Template Platform (designed and built by McDermott in 1947; courtesy McDermott International, Inc.)

Superior would have received even more kudos had their first well not been a dry hole. A small Midwest independent preempted them before they could bring on their second and successful well.

Credit usually goes to Kerr-McGee Corporation for ushering in the great and enduring oil bonanza that the Gulf of Mexico has provided. A tortuous struggle from 1945 to 1947 with financial and technical problems led K-M to build two bantam-sized platforms, one only 2700 sq ft, the other 3600, in the Ship Shoal Area, 10 miles off the Louisiana coast. On October 14, 1947, K-M snatched the brass ring ahead of the well-financed but ponderous Superior, beating them by eight months to first oil from a well out of land sight.

For design and installation, K-M used Brown & Root, McDermott's arch rival and a company anxious to establish a position in offshore work. Ironically, the design, a platform set on steel and wood piles, predated Superior's. However, their frugal but shrewd effort included the use of war surplus barges, air-sea rescue boats, and a landing ship tank (LST) for support vessels. They converted the 367-ft LST, to a drilling tender, adding a living quarters, a 35-ton crane, and winches for mooring. (*See* Fig. 1–5.)

When K-M completed their first well at almost 500 barrels per day, the combination platform and drilling tender captured the imagination of the industry, eclipsing Superior's technically superior—but 20 times larger—platform design. K-M had created a paradigm that lessened exploration risks by using fixed platforms of minimal size and mobile drilling tenders. In the event of a dry hole, the bulk of the investment—the tender and the topsides—could be redeployed to another site. LSTs moved to the top of companies' wish lists. Even Humble Oil, not always known for its quickstep, bought 19 LSTs the next year for conversion to drilling tenders.

Jackets and Templates

The unlikely term, *jacket*, came about when platform fabricators substituted steel frames for the wooden pilings supporting the decks. They manufactured the frames at onshore yards, towed them to the drilling location, and dropped them on the wellsite. To anchor the frame in place, the installers drove pilings, sometimes wood but later steel, through the legs of the frame, i.e., the *jackets*. The term was quickly extended to mean the entire structure that supported the platform. Later, as the frames grew to gorilla sizes, too large for a simple crane lift, the legs sometimes held ballasting tanks used to float the frames off the barges. *Templates* later became synonymous with jackets where pilings were driven using the legs as guides.

Fig. 1–5 Kerr-McGee's Platform in the Ship Shoal Area of the Gulf of Mexico with the Drilling Tender, *Frank Phillips*, a Converted Navy-surplus LST, Butted up to It in 1947 (Courtesy Kerr-McGee Corporation)

The boom had started, and numerous companies followed K-M's lead into the Gulf with platforms and tenders. But as they moved to deeper waters, they found that building even a small platform just to drill one or two exploratory wells became too expensive. Clearly they needed new concepts, and they found the next one in the Louisiana swamps.

Submersibles

Along came John T. Hayward, a marine engineer with the unlikely credentials of having supervised the first rotary-drilled oil well in Rumania in 1929. A partnership that included Seaboard Oil Company acquired a prospect in the Gulf of Mexico, without a clue as to the cost of drilling the six wells needed to explore it. In desperation, they turned to Hayward in 1948 for help. He mused about the barges that he had seen sitting on the bottom of the Louisiana swamps with drilling platforms welded on top. Simple linear scale-up for even 30- to 40-ft water depths would lead to 50-ft-high vessels that would drift away with only moderate tidal currents. Instead, he designed a totally submerged, conventional-sized barge with columns high enough to support a platform at a safe above-water distance with manageable freeboard and no drift. Pontoons on either side of the barge provided both stability and displacement control.

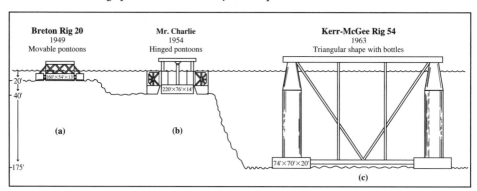

Fig. 1–6 Submersible Rigs: a) The *Breton Rig 20* in 1949, the First Submersible Used Offshore; b) Odeco's *Mr. Charlie*; c) Kerr-McGee's *Rig 54*, a Triangular Platform with Bottles at Each Apex (Rendering after Richard J. Howe)

Despite the initial, horrified response of his clients, who had difficulty with the concept of sinking a barge on purpose, he convinced them to build the prototype rig, the Breton Rig 20 (Fig. 1–6a). In early 1949, they used the rig to drill a half dozen exploration wells in the Gulf, moving 10 to 15 miles between each, drilling within a day or two of leaving the previous site.

The diciest step using *Breton Rig 20* came as the barge submerged—an untoward wave or current could flip it, especially in deeper water. Fortunately that didn't happen, and Kerr-McGee purchased the rig from the partnership and built two more like it. They worried over the stability, improving it on each edition, but after some near accidents, they contracted Odeco to build *Mr.*

Charlie (Fig. 1–6b), a submersible designed to handle the problem. They rigged the barge with pontoons at each of the long ends. It operated like an old man getting in a car—butt first. They ballasted one pontoon until that end of the barge sat on the bottom. (They still operated in only 20–40 ft of water.) With the stability ensured, they filled the other pontoon until the barge rested on the bottom, topside-up every time.

For another dozen years, companies tried design variations of these submersibles, pushing the water depths from scores to almost 200 ft. Some had outrigger hulls; some had large cylindrical tanks in the platform corners. Kerr-McGee, the leading company in submersibles, built the largest and last one, *Rig 54* (Fig. 1–6c), in 1962. The unusual-looking rig sported a triangular platform, bottle tanks at each apex 388 ft from each other, and could drill in 175 ft of water. Industry used the 30 submersibles built during this time until the 1990s. Meanwhile other keen minds worked on another innovation to reduce costs of exploration by eliminating the large amount of steel needed to fabricate barges, pontoons, tanks, and bottles.

Bootstrapping

In its chronically assertive fashion, the oil industry stole a concept that had long languished in the marine industry: the *jack-up*. Naval architects and civil engineers had been installing jack-up docks in remote locations around the world for decades, even using them at Normandy during the Allied invasion of Europe. At mid-century, Col. Leon B. DeLong built the most famous jack-up, a platform for radar towers 100 miles off Cape Cod in 60 ft of water. For this engineering feat, remarkable at the time, history immortalized the Colonel by thereafter referring to the concept as the DeLong design. The idea was simple. On a barge or other floatable, install tall cylinders (or

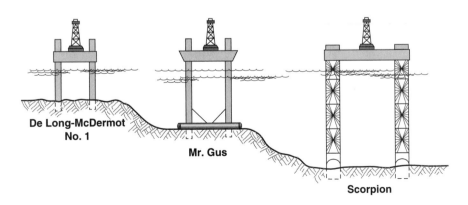

Fig. 1–7 Early Jack-up Rigs: the *DeLong-McDermott No.1*, Mr. Gus, and the *Scorpion*

caissons) around the perimeter. Float the barge to a site, and drop the caissons to the bottom like legs. Then jack the platform up the remaining length of the caissons as high above water as required.

In 1950, Magnolia Petroleum Company installed the first DeLong-design platform in the Gulf of Mexico. It stood on six caissons in 30 ft of water. Ironically, they used it as a permanent production platform, but McDermott Company followed with a mobile rig the next year, the *DeLong-McDermott No. 1.* (*See* Fig. 1–7.)

Not all efforts made one step forward. An embarrassing two steps back appeared in 1954 by the name of *Mr. Gus*, a design of Bethlehem Steel Company. *Mr. Gus* consisted of a barge, a platform above it, and four legs, all designed to operate in 100 ft of water. The platform stayed in place as the barge slid down the legs to the bottom—sort of a jack-*down*, not -*up*—to serve as a base for the platform.

Fig. 1–8 Aboard the *Scorpion*, First Jack-up to Use Rack and Pinion Drives (Courtesy George Bush Presidential Library)

On its initial installation in 50 ft of water, the barge tilted, breaking pilings and damaging two legs. Undaunted, Bethlehem took *Mr. Gus* to the yards and repaired the design problem. They sent the rig back out to the Gulf, whereupon it capsized in rough seas and sank off Padre Island, Texas, ending any residual interest in jack-downs.

In 1953, R. G. LeTourneau, who had made his fortune inventing modern earthmoving equipment, entered the offshore industry with a successful and enduring extension of the DeLong design. Rather than caissons around the perimeter, LeTourneau switched to steel truss-like legs. Borrowing from his experience with earth-moving equipment, he designed the lifting mechanism with rack and pinion drives and electric motors.

The established oil companies showed little interest in the strange configuration that LeTourneau proposed. It took an upstart, Zapata Offshore Company, to underwrite LeTourneau. On March 20, 1956, LeTourneau delivered the Scorpion to George H. Bush, Zapata's president and founder. (*See* Fig. 1–8.) This jack-up had six 152-ft legs in two triangular sets and an eight-million-pound platform. The dumbbell shape of the platform, an object of more than one disparaging remark, gained scant industry enthusiasm, but almost all jack-ups built after that used the rack and pinion lift design with electric drive.

Floaters

What could be better than adversity to stimulate creative juices? Oil companies long coveted the potentially prolific leases off the coast of Southern California. At the same time, Californians had never gotten over the unsightly morass at Summerland and countless other nearby aesthetic and environmental disasters. They raised strong objection to any additional permanent offshore platforms.

Continental, Union, Shell, and Superior Oil Companies formed a consortium, irreverently named the *CUSS* group, and commissioned the *Submarex*, a drilling ship. They converted yet another war surplus vessel, a patrol boat, by adding a drilling rig cantilevered over the port side amidships. In 1953, the rig drilled in depths of 30–400 ft, but vexing engineering problems quickly convinced the CUSS group that they and the *Submarex* were not yet ready for prime time and limited the operations to core sampling, not exploratory wells.

Still the CUSS group learned enough about stability, mooring, and drilling that they began design on *CUSS 1*, a purpose-built drilling vessel launched in 1961.

CUSS 1 had no self-propulsion. Tugs positioned it on a site; moorings held it in place. On board, a derrick perched above an access hole in the barge's center. Under that sat the key innovative mechanism, a birdcage structure on guide wires leading to a landing base on the ocean floor.

They started the drilling sequence with the birdcage on board. They ran surface pipe down through its center (Fig. 1–9a), almost to the ocean floor. A drillstring run through the surface pipe spudded the hole in the ocean floor. The surface pipe was sunk to a few feet. Blowout preventers were added to the birdcage, and lowered to the bottom. The pipe was cemented in place (Fig. 1–9b). Registry cones on the birdcage and the guide wires (Fig. 1–9c) facilitated landing additional equipment for subsequent drilling and completion. The design would outlast the century.

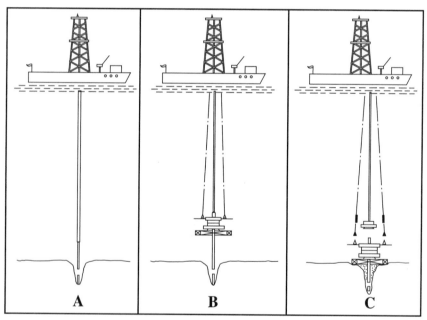

Fig. 1–9 The Drilling Sequence Used aboard the *CUSS 1*

CUSS 1 performed successfully in waters up to 350 ft, drilling core holes down to 6200 ft. At the same time, Standard Oil of California (Socal) and Brown & Root each experimented with derricks on barges similar to the *Submarex* and the *CUSS 1*, using them primarily for geological surveys. The Offshore Company made an obscure oil discovery in 1958 off the coast of Trinidad from a barge with a derrick. Still, most historians credit *CUSS 1* for starting a new class of exploratory drilling options: floating platforms.

Class Distinction

An obscure government office in New Orleans gave birth to the unlikely term *semisubmersible*. Before sailing, Shell had to apply to the Coast Guard for an operating license for the *Bluewater I*, the first of its class. Shell wanted to avoid using the term *ship*, on the application, lest the maritime unions claim jurisdiction. Bruce Collipp, the design engineer, explained to the local Coast Guard Commandant how the *Bluewater I* operated like a submersible but was only partially sunk. The commandant wrote on the licensing application, "Type Vessel: Semisubmersible," thereby naming the new class of drilling rigs.

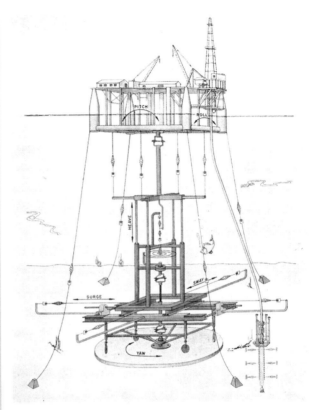

Fig. 1–10 Pioneer Bruce Collipp's DaVinciesque Diagram (Used to explain the six types of motion that ocean drillers have to cope with. This and his experience with the CUSS 1 led to his invention of the semisubmersible. Courtesy Bruce Collipp)

Much of the challenge around floating platforms centered on stability. Bruce Collipp, a naval architect and therefore an unlikely employee of an oil company, articulated all the movements that had to be accommodated while drilling from any floater—surge, sway, pitch, roll, heave, and yaw. His early diagram, shown in Figure 1–10 is legend.

While working at Shell, Collipp received inspiration while aboard an Odeco submersible. During a heavy seas episode, with the submersible under tow to a new location, the operator partially submerged the vessel to protect it from capsizing. Collipp noticed the immediate improvement in stability. He went on to design and patent the first large semisubmersible, the *Bluewater I.* This floating drilling platform started life as a bottle-type submersible, but Shell added additional ballast tanks and then only partially flooded the four bottles. The bottles then lay mostly beneath the water's surface. Their small profile at the water's surface reduced the effects of wave action and gave *Bluewater I* the stability that hulled vessels could not achieve at the time.

Drillers of exploratory wells in deeper water came to love the semisubmersible in its various sizes and shapes. Some unusual designs included Odeco's *Ocean Driller*, which had a V-shaped platform with multiple caissons; the *Sedco 135* had a triangular platform with bottles at each apex.

This is not to say that companies lost their enthusiasm for drilling from self-propelled, hulled vessels. In parallel development, purpose-built drillships went into service. In 1962, Sedco built the *Eureka* for Shell Oil. (See Fig. 1–11.) To deal with positioning, the *Eureka* had port and starboard propellers extending from the bottom of the ship. The propellers could be rotated 360° to move the ship in any direction.

Up to that time, drilling from barges necessitated dropping buoys around the wellhead to give a clue about the appropriate position. The barges had to be anchored in four directions; anchor lines had to be continually winched in and out to maintain the correct position over the well. The

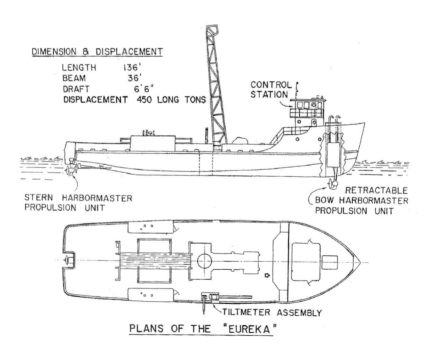

Fig. 1–11 The Original 1961 Drawings by H. L. Shatto and J. R. Dozier of the Drillship *Eureka*; Bow and Stern Thrusters Rotated 360° to Maintain the Ship's Position (Courtesy Howard Shatto and Shell Oil Company)

Shatto's Tale

Howard Shatto, a pioneer in drillship stability, tells the story of maneuvering vessels to place a 15,000-ton platform section onto a site in the Gulf of Mexico. In 1975, his employer hired a company to do the lowering using anchored barges. Shatto asked the marine captain how long it might take, once the platform was floated between the barges, to move it, say, 50 ft to center it over the selected site. The captain estimated that manipulating the 12 anchor lines through their respective winches would take about 12 hours. Shatto knew two tide changes and their currents would push the platform around even more than the required 50 ft. After some consideration, he pointed out that to move laterally 50 ft, the captain need only let out 50 ft from one restraining anchor line, take in 50 from its opposite, and adjust the other 8 lines by the cosines of their angles.

The captain and even Shatto's colleagues mounted vigorous, if not rigorous, objections based on both the complicated mathematics of catenary–shaped anchor lines and 20 centuries of marine lore. In the end, they agreed to try it and found that, indeed, a foot of anchor line, or its cosine-adjusted equivalent, gives a foot of movement, and in about three minutes. Shatto revealed the logic behind his insight only after selling the calculation on magnetic strips for hand-held calculators to dozens of companies for $10,000 a copy. (Shatto's aha!: at the bottom of the mooring line, more than 100 ft of mooring line and anchor chain lies between the anchor and the point where the line starts its catenary-shaped rise to the vessel. Winching in one foot lifts one foot off the bottom and leaves the catenary shape unchanged.)

Eureka needed no anchors, but it did have a positioning device tied to the ocean floor, a thin wire that ran up through an on-board *tiltmeter*. This mechanical device measured the angle of the wire and calculated the position of the ship relative to the wellhead. Operators then used a joystick to engage the forward and aft thrusters. Experience indicated the joysticks were about as stable as an arcade road-racing game, and with about the same results. The designers quickly replaced them with automatic electro-mechanical devices (using vacuum tubes!) that performed much better.

Shell, the *Eureka* owner, limited the ship's use to drilling core samples, deeming this ship, like the *Submarex* and the *CUSS 1*, too experimental to drill exploratory wells. More than 10 years passed before a purpose-built, dynamically positioned drillship, *SEDCO 445*, appeared in 1971, ready to drill exploration wells. Rotating fore and aft thrusters seemed clever enough during the *Eureka*'s design, but at sea, the operators literally wore them out as they swung them to and fro to maintain position and with voluminous use of fuel. The *SEDCO 445* had fixed thrusters, 11 along the port and starboard that gave lateral and heading control. The main screws provided fore and aft positioning. This simpler, more durable design became the standard for subsequent drillships.

Honing Tools of the Trade

Whether the drilling derrick sat on a semisubmersible or a dynamically positioned drillship, the driller continually asked, "Are we over the wellhead?" In shallow waters, for a while some companies anchored their vessels and continued to use variations of the Eureka's guide wires. In deeper waters, where they could tolerate more movement, they moored buoys in a circle as guides, sometimes using a tiltmeter in parallel. Then they worked the anchor lines to move the vessel. Aboard the *Bluewater I*, substituting wire rope for anchor chain to make winching and positioning easier came as a welcome but forehead-slapping innovation in 1961.

Much deeper water made that method impractical because of the long mooring lines. The *SEDCO 445* used an acoustic positioning system. Pingers on the wellhead sent signals to the ship's hydrophones. A few trigonometric calculations gave the position, although the error due to uneven water speed could be plus or minus 1–2% of the water depth. (3000-ft depths could give a 60-ft error!) Later operators switched to placing four transponders at some distance around the wellhead. The ship sent a signal, the transponders sent it back, and onboard computers triangulated (quadrangulated?), reducing the error to a more tolerable one-half percent.

A big leap in position determination came in the 1980s when enough satellites whizzed around the globe to give continuous line-of-sight coverage. GPS, the Global Positioning System, eventually let ships know where they were within a few feet.

Divers and ROVs

From the beginning, drilling at sea called for subsea assistance—to locate wellheads, to make connections, to set platforms, to do inspections, and for many other tasks. Early efforts used divers who could operate efficiently down to about 100 ft. Pressures beyond that could turn divers into blithering idiots with the attention span of a fly on a fresh cow patty.

Positioning by Chance

By 1986, the U.S. Government had enough satellites up to provide continuous signals in the Gulf of Mexico. But foreign enemies could use the Global Positioning System (GPS) to lob missiles into America just as easily as the oil industry could use it to dynamically position their drillships. In the name of national security, the government diddled with the signal to ensure the bad guys would have inaccurate launch positioning. That also rendered the signal useless to the drillers.

Along came John Chance who figured he knew where he stood, so he could continuously calculate the diddle correction. He did and he signed up the drillship companies and continuously transmitted to them the offsetting error, allowing them accurate GPS use.

Later Chance's company, Starfix, switched to undiddled commercial satellites (which still needed some correction) and continued to provide tracking service to a growing fleet of dynamically positioned vessels.

The U.S. Navy had discovered that using a mix of oxygen and helium instead of air could push divers' limits down past 200 ft. Helium replaced the nitrogen content of the air, eliminating the culprit that easily penetrates brain tissue, inducing narcosis, and causing divers to act like 2-year-olds. Of course, the helium made them sound like Donald Duck, but at least they knew what they were doing. In 1960 in the Gulf of Mexico, Shell sponsored the first use of oxygen/helium in offshore exploration.

Still, even using helium, diving required long and expensive periods of diver decompression. Industry needed a nonbreathing, underwater assistant and began to experiment with robots. *Mobot*, the first remote operated vehicle (ROV) used to complete an offshore well, went into service in January 1962 for Shell. Long before George Lucas conceived R2D2 and C3PO, this elegant little robot (Fig. 1–12) had four distinct features:

- free-swimming self propulsion
- on-board sonar that could find a wellhead
- a television camera that could see it
- a socket wrench that could connect a Christmas tree or a blowout preventer

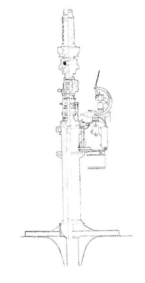

Fig. 1–12 Original 1962 Drawing of the *Mobot* Clinging to a Wellhead (The profile shows, top to bottom, the tether, television camera, sonar apparatus, ratchet wrench, and wheels that permitted it to circle the wellhead; courtesy of H.L. Shatto and Shell Oil Company)

Over the next 10 years, *Mobot* successfully completed six subsea wells in the Molino Field off California, a discovery well in Cook Inlet, Alaska, and 18 more exploration wells up and down the U.S. West Coast, all without the assistance of divers.

During that same time, other ROVs entered service with a variety of capabilities—articulated arms, grabbing devices, suction cups, high-pressure jets for cleaning, and other tools. Operators controlled them aboard the drilling vessel with joysticks, television receivers, and even early versions of virtual reality apparatus.

In parallel, the diving industry, led by Taylor Diving and Salvage Company, mastered "saturation diving," which permitted divers to stay on site in hundreds of feet of water indefinitely, using pressurized habitats and carefully monitored decompression. In a 1970 experiment, five Taylor divers worked 18 days from a pressurized vessel at simulated depths of 1000 ft. In the ensuing decade, Taylor's crew broke successive commercial records, culminating in a pipeline job in 1978 for Norsk Hydro. Taylor divers welded two sections of 36-in. diameter pipe in 1036 ft of water offshore from Western Scotland.

After that, the use of ROVs and diving converged, with divers handling the fine motor skill tasks and the somewhat clumsier ROVs doing the heavy-duty, surveillance, and some specialty work.

Lift Power

Transporting and launching prefabricated production platforms required newly designed barges to haul them and mobile heavy lifting equipment to get them in the water and correctly positioned. (*See* Fig. 1–13.) Over the last half of the 20th century, floating crane capacity increased dramatically (Table 1–1) as industry pushed into deeper and rougher waters and jackets grew larger and heavier. Following the lead of the innovative and entrepreneurial P. S. Heerema, other companies upgraded their crane capacities, or built new crane vessels.

Geology, Geophysics,
and Other Obscure Sciences

Ask a *geologist* a question about the offshore that includes the word *history*, and you'll likely get a long answer that begins not a hundred years ago but a hundred million. You'll hear that the Gulf of Mexico, for instance, became so rich in hydrocarbons because ancient rivers, ancestors of the Mississippi, deposited a continent's worth of organic material, shales, and sands in long

Fig. 1–13 Twin Cranes Lifting a Jacket into Place (Courtesy Shell International, Ltd.)

Table 1–1 Heavy Lifting Milestones

1948	75-ton crane lift of Superior's jacket at the Creole field
1962	300-ton crane on Heerema's ship, *Global Adventurer*, into service
1968	800-ton crane on Santa Fe's *Choctaw*, a column stabilized catamaran
1972	2000-ton crane on Heerema's ship, *Champion*, into Amoco's service in Suez
1974	2000-ton crane on Heerema's ship, *Thor*, into BP's service at the Forties field in the North Sea
1976	3000-ton crane on Heerema's ship, *Odin*, installs the platform on Shell's Brent Alpha jacket
1977	2000- and 3000-ton cranes installed on Heerema's ships, *Balder* and *Hermod*
1986	*Balder* and *Hermod* crane capacities altered to 4000 and 5000 tons
	Twin 6000-ton cranes installed on McDermott's semi submersible
	Twin 7000-ton cranes installed on Microperi's semi submersible

By 2000, the *Microperi 7000* had been acquired by Saipem who upgraded it to 7840 tons per crane and Heerema had upgraded their *DCF Thialf* with two cranes of 7810 tons each.

fairways running into the present Gulf Coast. The weight of the sediments created enough pressure and temperature to cook the organic matter, some into oil, some into gas. Along the coastline, wandering landmasses stranded seawater, which evaporated and left huge sheets and pillows of salt. The shale provided the source rock, sands provided the reservoirs, and cap rock plus salt trapped the hydrocarbons.

Incoming!

On both sides of the trenches of World War I, groups of mathematicians, physicists, and engineers used acoustic equipment to plot the locations of enemy artillery. They took readings at three or more locations to triangulate on enemy guns. In the 1920s, some of these same people came to America and developed the seismic refraction and later seismic reflection techniques, and with that, they founded some of the earliest geophysical companies. One Frenchman had a name that now speaks for itself, Captain Conrad Schlumberger. Another, a German named Ludger Mintrop, founded Seismos, Limited, a firm ultimately absorbed by Schlumberger to form the core of their seismic services subsidiary.

Fig. 1–14 Early Offshore Seismic Collection (Courtesy Western Geco)

The story then shifts to somewhat later, about 1920, when geologists began to realize the similarities between the onshore Gulf Coast and the Continental Shelf. After all, it was only around 1912 that exploration companies began to hire geologists. That year marked the first recorded discovery, the Cushing, Oklahoma Field, directly resulting from a geologic survey.

Ask a *geophysicist* the same question, and the history starts less than a hundred years ago. In 1924, Amerada Oil discovered the Nash salt dome in Brazoria County, Texas, using an early mapping tool, the torsion balance. Two years later, Amerada completed an oil well there, making the Nash field the first oilfield credited to any geophysical method.

Seismography developed at the same time. During its early stages in the shallow inland lakes and swamps and in the offshore, seismic crews planted geophones by hand, with locations measured by land sight. Recording apparatus sat on raft-like vessels. (*See* Fig. 1–14.) Geophysicists at Teledyne credited their full-scale marine survey in 1934 for the discovery of Superior's Creole Field.

Within 10 years, self-contained, 60-ft boats towed hydrophone cables into place, backed up a bit to let them settle, than radioed the shooting boat to drop a dynamite charge and back off. Blessedly, the seismic waves they captured, despite having to travel through scores, then hundreds and eventually thousands of feet of water, behaved no differently than onshore waves.

Seeing the Forrest

In 1967, a young geologist noticed a curious pattern in the seismic data shot in preparation for lease sales offshore Louisiana—an abrupt zone of low velocity reflections. The same phenomenon turned up a year later on a Bay Marchand prospect. When his company drilled wells through those zones shortly thereafter, they confirmed commercial gas accumulations in each.

For another year, he assembled more evidence, debated with skeptical colleagues, and pestered his management until his vice-president agreed to see him and review his story. The geologist, Mike Forrest, convinced his management about the remarkable power of *bright spots*, a name coined during coffee room debates. They could use seismic data to do more than just geologic mapping. They could directly identify natural gas accumulations.

For the next few years, he joined teams of geologists, petrophysicists, geophysicists, and computer scientists correlating *bright spots* with every other piece of evidence they could. As confidence grew, Forrest doggedly spurred his company, Shell, to win leases and ultimately prove with the drilling bit hundreds of millions of barrels of hydrocarbon in the Gulf of Mexico. Finding the 300 million barrel reserve on Prospect Cognac in 1975 using *bright spots* was only a prelude to the giant fields discovered in the deepwater in the next decade.

Soon neutrally buoyant tubes with hydrophones allowed the boats to record while underway. Eventually the geoservices companies also bowed to the environmentalists and fishing industry who were understandably upset at the sight of a seascape of dead fish after a seismic shot. By the 1970s, most seismic boats towed a steel cylinder charged with compressed air that, on release, sent a pressure wave as a signal, with as good a result as an explosive.

Seismic data interpreters were among the first disciplines to fully exploit contemporary computing power. In 1958, Geophysical Services, Inc. (GSI) fired up the first digital computer wholly dedicated to seismic processing. With that, paper recording gave way to analogue recording and eventually to digital recording.

At the annual meeting of the Society of Exploration Geophysicists (SEG) in 1970, Exxon Production Research presented the results of seven years' efforts, their breakthrough on 3D seismic. In another two decades, geophysicists sat at computer workstations and manipulated data, relaying it to "Spielbergesque" display rooms, where they sat with geologists surrounded by brilliantly colored displays of the subsurface. All this advanced the discovery and appraisal process and reduced the risk of dry holes.

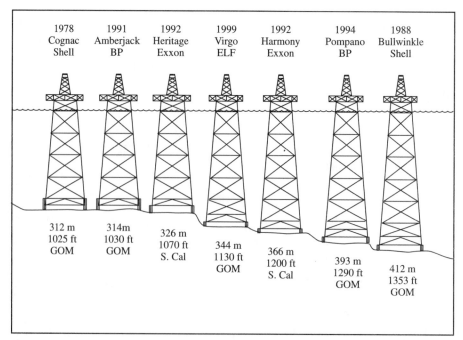

Fig. 1–15 Fixed Platforms by Installation Year and Installing Company: Two Decades at the Limit

Permanence

As the geologists and geophysicists demystified the subsurface and mobile rigs drilled exploratory wells further from the shoreline, the demand grew for permanent, durable production facilities in ever-deeper water. Building on-site at sea became out of the question, and fabrication yards sprang up along the Gulf Coast.

Rediscovering the success of Superior in 1947, operators learned to love prefabricated jackets. The simple concept involved fabrication onshore, barging (or later floating) the jacket to the site, lifting it off the barge, and lowering it on the target. Driven piles made from open-ended steel tubes, eventually as large as 8 ft in diameter, held the jacket in place. Then came the topsides, lifted in toto or in parts from the transporting barges and fitted to the stubs of the jacket sticking out of the water.

No one would call progress in jacket development dramatic—until the 1970s and 1980s. For 30 years, engineers had worked to increase strength and decrease weight and drag, keeping the cost

Fig. 1–16 The *Bullwinkle Platform* Being Towed to Sea (Courtesy Shell International, Ltd.)

the borehole from collapsing, and that simple morsel of technology made his well a producer and gave his reputation immortality.

Not far away, the next year J. C. Rathbone drilled into a small hill at Burning Springs in West Virginia and brought in a well that produced 10 times the rate of Drake's well. After that, a growing horde of wildcatters took note that oil seeps often occurred on natural surface bulges, later called anticlines. These natural formations, like the one at Spindletop, provided good geological mechanisms for trapping accumulations of oil and gas. Starting with this knowledge, waves of entrepreneurs, engineers, scientists, financiers, and fortune-seekers created massive practical and intellectual breakthroughs and pushed up the onshore learning curve for a century. Figure 1–18 shows only a few of the milestones, which continued well beyond the first ventures into the offshore.

The dynamic development of onshore exploration and production technology provided only the necessary backdrop. But moving offshore in the 20th century needed the new paradigms and the new breed of oilmen to get from Summerland to Bullwinkle, the second wave shown in Figure 1–18.

But to some, as the 20th century drew to an end, even this learning curve had played itself out. The offshore industry needed a jumpstart.

2

Letting Go of the Past

It is the image of the ungraspable

. . . this is the key to it all

From *Moby Dick,* Herman Melville (1819–1891)

The Confusion of the 1980s

Nothing but promise appeared on oilmen's horizons in the early 1980s. OPEC provided a dizzyingly high oil price, at one point more than $40 per barrel. Rigs operating in the Gulf of Mexico increased to 231 in 1981, double that of 1975, and almost triple the count in the early 1970s (Fig. 2–1).

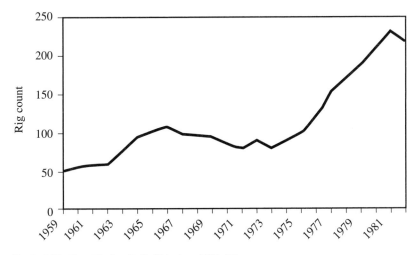

Fig. 2–1 Rig Count in the Gulf of Mexico, 1959–82

Worried about energy security, the U.S. Government stopped its piecemeal offerings of tracts and held its first Gulf of Mexico area-wide lease auction in 1983. They included broad expanses of tracts both on the Outer Continental Shelf plus ones extending into waters of 2000–7500 ft depth.

Advances in drilling technology, especially dynamically positioned drillships, anchored semisubmersibles, subsea completions, and the use of 3D seismic, a costly but technologically superior advance over 2D seismic, allowed drillers to test these deep waters. Rumors of a succession of elephant-sized discoveries abounded (Table 2–1), huge in comparison to discoveries on the Continental Shelf. Conoco, Shell, British Petroleum, Exxon, and Oryx (who ultimately disappeared into Kerr-McGee) pioneered exploration success in the new province.

Table 2–1. Significant Deepwater Discoveries in the 1980s

Field	Volume Million barrels	Depth, ft	Company	Year
Joliet	65	1724	Conoco	1981
Pompano	163	1436	BP	1981
Tahoe	71	1391	Shell	1984
Popeye	85	2065	Shell/BP/Mobil	1985
Ram-Powell	379	3243	Shell/Amoco/Exxon	1985
Mensa	116	5276	Shell	1986
Auger	386	2260	Shell	1986
Neptune/Thor	108	1864	Oryx/Exxon	1987
Mars	538	2960	Shell/BP	1988

In the mid-80s, ominously dark clouds floated over the Gulf of Mexico. OPEC learned that with $34 oil, they had priced themselves out of many markets. Consumers found unprecedented ways not to use oil, in their cars, industrial plants, and buildings. The oil price softened to $28 dollars in 1983 and collapsed to $10 in 1986.

Coincidentally, exploration success in the Gulf of Mexico Continental Shelf, the region in less than 1000 ft of water, began to show signs of increasing age. The average size of discoveries declined by half from the previous decade to about 24 million barrels. (*See* Fig. 2–2.) Despite the huge increases in rig activity, after 25 years of double-digit growth rates (with only a short pause in the 70s), the production of oil and gas from the Gulf flattened out like the Cumberland Plateau (Fig. 2–3).

Oilmen searched in their files for the U.S. Department of Interior report from a decade earlier, which predicted that essentially all the potentially productive blocks in waters up to 600 ft

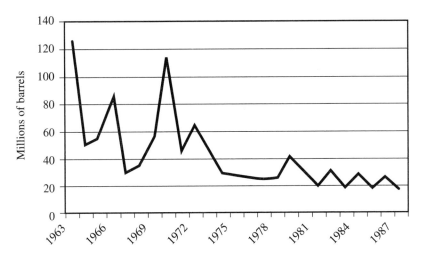

Fig. 2–2 Average Size of Fields Discovered in the Gulf of Mexico, by Year in Which They Were Discovered (U.S. Minerals Management Service)

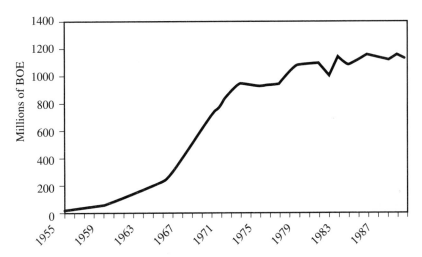

Fig. 2–3 Oil and Gas Production in the Gulf of Mexico in Barrels of Oil Equivalent (U.S. Minerals Management Service)

would have been already leased and development would have been completed by 1985.[1] Even more depressing, they began to refer to the Gulf of Mexico Continental Shelf as the "Dead Sea."

With the exception of Prudhoe Bay, Alaska, the notable lack of success for 10 years by the major oil companies elsewhere in North America disheartened the most stalwart oilmen. Huge exploratory expenditures in the Norton and Navarin Basins in offshore Alaska and in Georges

Banks, offshore East Coast, yielded no commercial hydrocarbons; discoveries on the entire onshore Lower 48 were puny; and environmental restrictions eliminated any new plays in the West Coast offshore. By 1986, the oil company annual reports began to announce new exploration strategies—a shift from the United States to the onshore and offshore areas of other continents, Asia, Africa, Australia, and South America.

That is not to say momentum didn't sustain some progress in the Gulf of Mexico. In 1984, Placid Oil, H. L. Hunt's intrepid independent company, announced a large discovery in 1554 ft of water. They immediately undertook development of the field with subsurface wellheads producing back to a floating platform, a first of its kind in the Gulf. At the same time, Conoco mulled over developing its 1981 discovery, *Joliet*, in 1760 ft. Conoco had developed some expertise in the North Sea using a floating tension leg platform (TLP) to develop their Hutton field. Emboldened by that success, they installed the first TLP in the Gulf of Mexico in 1989.

Disconcerting signals

Then the petals fell off the technological bloom. Having installed *Bullwinkle* (a conventional fixed base platform) in 1354 ft of water in 1988, Shell reached the economic limit for this configuration. Using more than its 54,000 tons of steel structure and pilings would strain the profitability of any conceivable project in deeper water. In addition, the pyramidal base measured some 400 by 480 ft, barely manageable dimensions.

While Placid's and Conoco's approaches were less capital-intensive engineering successes, it became apparent they had misjudged the geology of their fields. Poor performance of their wells led Placid to abandon their investment. Conoco's *Joliet* limped along, cash positive but with below-investment-grade performance.

At this point, the U.S. offshore industry faced a crisis of confusion:

• volatile and mediocre oil and gas prices
• unpromising potential in the Gulf of Mexico Continental Shelf
• apparent brighter opportunities in foreign plays
• billions of barrels of hydrocarbons already discovered but not developed in the deepwater
• uncertain deepwater reservoir performance that put development investment at risk
• yet more Federal lease sales scheduled

Oilmen pondered, *What should we do?*

Turning the key

Coincidental and fortunate circumstances, only vaguely related to each other, occurred about that time. In 1974, Petrobras, the national oil company of Brazil, began exploring with modest success the Campos Basin off their northeast coast. Brazil, like many other countries, worried over energy security. Their oil imports ran at 70–80% of consumption. In the early 1980s, the Brazilian government challenged Petrobras to substantially reduce their dependence on foreign oil.

At their first commercially significant find in 1974, the Garoupa Field in 394 ft of water (Fig. 2–4), Petrobras had used a conventional fixed-base platform. After that, in a burst of productivity and originality, they discovered *Bonita, Enchova, Piraúna, Marimbá, Albacora,* and *Barracuda* and produced them from subsea wells first into temporary and then permanent floating production systems (FPS's). These FPS's, moored above the producing fields, provided a transit point for shuttle tankers to load the crude oil. Using FPS's avoided the construction time associated with

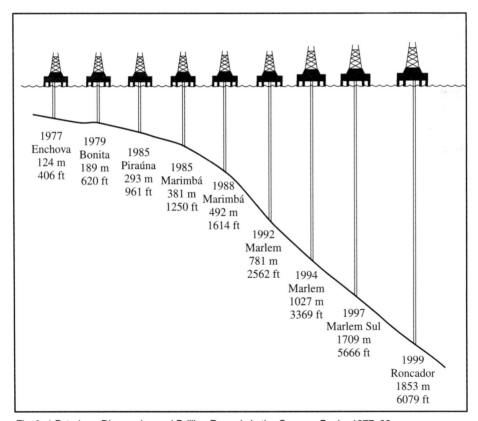

Fig. 2–4 Petrobras Discoveries and Drilling Records in the Campos Basin, 1977–99

both fixed-base platforms and subsea pipelines to shore and shortened the discovery-to-production-of-first-oil time from a typical nine-year span to five to seven.

In 1985, they found the *Marlim* field and in 1987 the *Marlim Sul* field, both two to three times larger than the elephantine discoveries in the Gulf of Mexico. *Marlim* sat under 2500 ft of water, but the *Marlim Sul* reservoirs lay beyond the continental shelf in depths from 2500 to 6300 ft. Exploiting all they had learned from multiple applications of their Early Production System, Petrobras tapped *Marlim Sul* in 1994 with a subsea completion producing into the *FPSO II*, a converted tanker.

To the delight of the Petrobras staff, the wells proved so prolific, producing upwards of 12,000 barrels per day, they justified accelerating their program of subsea development on Marlim and Marlim Sul using floating production, storage, and offloading systems (FPSOs) and FPSs. By 2000, they had 29 active FPS's and FPSOs in place in the Campos Basin.

Meanwhile, back in the United States, another company labored under the similar competitive constraints. Shell Oil Company was a wholly owned subsidiary of Anglo-Dutch giant, Royal/Dutch Shell, but they operated autonomously at the time. However, their parents already covered most of the attractive prospects outside the United States, so Shell Oil chose not to divert substantial spending to foreign plays.

Like Petrobras, they turned to the offshore and embarked with single-minded purpose on an *elephant hunt*—discoveries whose size would warrant spending hundreds of millions of dollars in development expenditures. After a discouraging series of exploration failures off Alaska and the U.S. East Coast, they focused with abounding success on the waters beyond the Gulf of Mexico's Continental Shelf. Shell chose to develop their first, the *Auger* field, which lay beneath 2860 ft of water, using a TLP. This concept had already been demonstrated by Conoco in the North Sea and in the Gulf of Mexico. But even with Conoco's and Placid's deepwater experience in mind, the subsurface risk—well productivity—remained.

Three shrewd moves buoyed their success. First, at unprecedented expense, Shell ran 3D seismic studies back and forth over the *Auger* prospect. That increased the reliability of their estimates of how much hydrocarbons the reservoirs might contain.

The second move took place not at *Auger*, but at *Bullwinkle*. There, Shell's production and reservoir engineers convinced their management to open the wells beyond any comparable shallow water producing rates. They had studied turbidite reservoirs around the world. They noted that these layers of sandstone had been deposited by turbid currents, which in certain cases had washed away the finer grains of sand as they lay down. That made these sandstone formations more porous and permeable, qualities high on a reservoir engineer's wish-list. The engineers convinced themselves that the turbidite sands below *Bullwinkle* would behave differently from the deltaic sands common to the adjacent Gulf of Mexico Continental Shelf.

If they misjudged, the rush of fluids could cause sand from the reservoir to plug up around the wellbore, with irreversible harm to the wells—and their careers. As they opened the valves for a few tense hours, the production rate went from 3500 to 7000 barrels per day. Bottom hole pressures remained constant, a crucial sign. No other bad symptoms of well behavior occurred. The experiment confirmed that reservoirs in the turbidite sands of the deepwater could produce at several times the rates the industry had grown to expect. They immediately reworked their *Auger* development plans, counting on producing 8000–10,000 barrels per day from each well, which happily they did. By having to drill only half the number of wells to drain the reservoir, they reduced *Auger's* cost by hundreds of millions of dollars—and lowered the worry about falling oil prices by at least half.

Third, like Petrobras, they accelerated development and reduced the interval to first oil. In this case, however, they predrilled the wells from readily available semisubmersible drilling rigs, the type normally used for exploration. Once the TLP was built and anchored in place, they completed the predrilled wells and started production.

Still, *Auger* took 10 years from wildcat discovery to full production. Oil companies, in cooperation with their contractors, set their sights on cutting the cycle time to less than five years.

Closing the door

With these bold steps, the doors to the past closed. Industry now had in hand the keys to unlock deepwater development:

- They realized the high productivity of turbidite sands, both in rates and ultimate recovery from the reservoirs.

- They could reduce uncertainty about reservoir configuration and the size of the hydrocarbons reserves through the use of 3D seismic.

- They could lower cost, innovative development systems—TLPs, FPS's, FPSOs, and later compliant towers, spars, and long-reach tiebacks to existing facilities.

• They could compress the time from discovery to first production through rapid approval, pre-drilled wells and other parallel activities, and quick construction of simpler facilities.

The third wave had begun (Fig. 2–5).

• • • • • • •

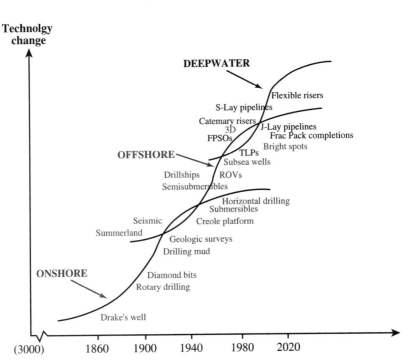

Fig. 2–5 Deepwater—The Third Wave in the Evolution of Oil and Gas Exploration

From time to time, Shell and Petrobras take victory laps for having pioneered the deepwater—and well they should. But scores of other enterprises contributed technology and technique to tapping hydrocarbons in the deepwater—drillers, mud companies, cementing services, fabricators, geoservice and seismic companies, maritime services, and more, not to mention the other oil companies. It takes another eight chapters to do them justice.

Notes

1. U.S. Department of Interior, Bureau of Mines, *Offshore Petroleum Studies Estimated Availability of Hydrocarbons to a Water Depth of 600 ft from Federal Offshore Louisiana and Texas through 1985* (December 1973)

3

Exploring the Deepwater

We must learn to explore all the options

and possibilities that confront us

in a complex and rapidly changing world.

Speech in the U.S. Senate

J. William Fulbright, 1905–1995

The oil business starts with exploration, and the exploration geologist, a scientist trained in the study of the earth, carries the baton through the first lap. He or she takes the responsibility of identifying oil or gas deposits of sufficient size to be drilled, developed, and produced. In this phase, the exploration geologist collaborates with a collection of scientists, engineers, and professionals in the process shown in Figure 3–1.

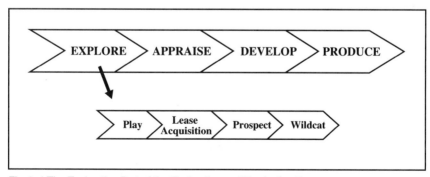

Fig. 3–1 The Exploration Part of the Exploration and Production Process

Developing a play

The exploration geologist starts with a region or a basin such as the Gulf of Mexico or the Campos Basin and develops a subsurface geological story whose plot has four dramatic episodes, which can be found in Figure 3–2.

- The source—a deep rock layer where enough organic matter, microscopic animal or plant organisms that lived millions or hundreds of millions of years ago, deposited in rich quantity; with the heat and pressure from the layers above, it cooked into forms of petroleum—oil and gas.

- The migration—gravity plus water intrusion, *geologic* forces, or a combination of the two, caused the oil and gas to migrate through the earth.

- The reservoir—adequately *porous* and *permeable* rock or sands that were available to receive the migrating oil and gas.

- The trap—where large enough accumulations are still confined in the reservoir by some *lithological* or *geological* phenomenon (the presence of shale, salt, or faults).

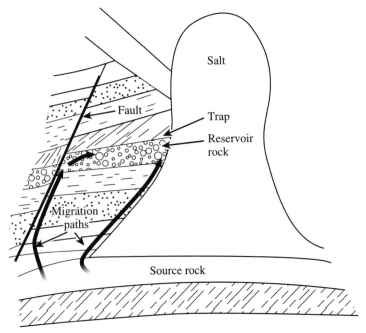

Fig. 3–2 Reservoir Rock, Trap, Sealing Mechanism, and the Migration of Hydrocarbons

> **Definitions Related to Petroleum Formation**
>
> • Lithology — The nature of the mineral content, grain size, texture, and color of rocks.
>
> • Geology — The study of the Earth: its history, structure, composition, life forms, and the processes that continue to change it.
>
> • Porosity (porous) — The percentage of pore volume or void space within rock that can contain fluids.
>
> • Permeability (permeable) — A measure of the ease with which fluids can flow through a rock by moving around the particles or grains that make up the rock.

This combination of source rock, migration, reservoir rock, and trapping mechanism becomes the basis for a *play*. With a technically competent story, the exploration geologist can convince the corporate decision makers to commit real money to the next several steps in the exploration process.

How do they know what's down there? The exploration geologists draw data from a rich body of knowledge and a broad slate of specialists.

- Geologists have developed a modern theory of plate tectonics. They say that the earth's surface is constantly moving, sliding along on 60-mile thick plates. As these plates drift around the globe, in some places they rub against each other, causing earthquakes, opening up fissures to the deep, molten magma, otherwise known as volcanoes, squeezing large pieces of real estate up like toothpaste out a tube, otherwise known as mountain ranges, and generally deforming the entire landscape.

- Geochemists help the geologist identify the composition of underground formations, in search of adequate source rock where the organic material might have resided. Good source rocks are typically limey material and shales and might contain 1% or more organic matter. Some fortuitous efforts find extremely rich source rock with as much as 10%.

- Geophysicists have become the eyes and ears of the exploration geologists, particularly during the preliminary evaluation stage, by contributing seismic data. The elaborate collection of data they bring to the party forms the basis for detailed subsurface models tied to the geological history.

Acquisition

Most enterprises that have any commitment to customer satisfaction and a sense of competition subscribe to the salesman's mantra, "Nothing happens until I make a sale." But E&P

companies in general have no preoccupation with that commitment. They sell their oil and gas into a faceless commodity marketplace. That is not to say they don't compete with other companies in the marketplace. They do, but mostly for the rights to drill for oil and gas. And that leads to the explorers' mantra, "Nothing happens until I get access."

At the early stage of developing a play, a company probably does not own or otherwise have the right to drill a well to confirm the existence of hydrocarbons in the alleged reservoir. In offshore waters throughout the world, the national governments own both the surface and subsurface rights and set the rules by which E&P companies can acquire them.

In the Gulf of Mexico waters that border the United States, the federal government administers control through the Mineral Management Service (MMS) an agency of the Department of the Interior. Annually, the MMS makes available to oil companies the opportunity to acquire the rights to explore and produce from wide areas of the offshore, including the deepwater.

What's in a Name?

Bullwinkle? Cognac? Where did these improbable names come from? Working on a play (an exploration project) is a cause for super-secrecy in an oil company. For that reason, some use code words when deciding how much to offer for a lease—the rights to drill on a particular spot—particularly in the offshore. A federal offshore lease sale may involve 10 or more prospects for a company, so some companies use a list of like-minded names for that sale. The names stick with the successful projects. That's why *Popeye* and *Bullwinkle* could carry over from one sale, and *Cognac* and *Neptune/Thor* from two others.

E&P companies submit sealed bids in competition with each other. Minimum bid for an offshore lease (a block 3 miles square or 5760 acres) is typically $400,000. Some winning bids, or bonuses, have run in the tens of millions of dollars. Besides the bonus, a company agrees to pay a royalty of 12.5–16% of the value of the oil and gas produced by the company over the life of the lease. Leases generally run for an initial term of 5–10 years, increasing with the depth of the water—and therefore roughly with the cost and risk involved in drilling a well on the lease.

(For the onshore United States and Canada, unlike anywhere else in the world, both private parties and governments can own the mineral rights to oil and gas. An oil company interested in exploring and producing on lands involving private owners can negotiate directly with him or her without involving the government.)

If the company drills a successful well on the lease, they retain the rights to the lease as long as they continue to produce oil or gas. If the company does not drill a well on the lease within the initial lease period, they lose the lease, which reverts to the MMS. It goes without saying that the government keeps the original bonus money and may re-lease the block to another company.

The region of the Gulf of Mexico that borders Mexico is administered by the Mexican Government, who historically has granted Pemex, the National Oil Company, the exclusive right to explore and develop those opportunities. In other international offshore plays, including in the deepwater off Brazil and the West African countries, the national governments typically have E&P companies submit more elaborate proposals for the rights to explore and produce, in this case more cavalierly called *concessions*. The proposal stipulates upfront fees (as in bonus), explicit expenditure plans for seismic data acquisition, and the expected number of exploratory wells in the early years of the concession. International concessions are typically 10–100 times the size of a Gulf of Mexico lease.

Many countries with less commitment to capitalism than the United States also leverage their monopolistic ownership position by having their national oil company own an interest in the concessions. That enables both financial opportunity for the national oil company plus technology transfer to the country.

Companies and countries negotiate the details of these arrangements upfront. In both the United States and internationally, landmen, professionals often with legal or business background, lead the negotiations.

Identifying the Prospect

Acquisition of the lease or concession triggers another round of effort by the exploration geologist, in concert with his team. They put new effort into understanding the geology. In some cases, more likely in shallow water than deep, the existing seismic data is enough to identify specific *prospects* with high enough probability of success to warrant drilling a *wildcat*, a well in an untested area. However, in most offshore cases, especially in the deepwater, hiring a geophysical service company to acquire additional seismic data makes sense. Acquiring, processing, and interpreting the data may take 12–18 months. All this is foreplay to drilling a wildcat well that might cost $25–100 million. The exploration geologist keeps in mind Darryl Royal's famous quote about forward passes. "Three things can happen when you throw a football. Only one of them is good." The same principle holds for drilling a well. Indeed, the results of additional seismic studies can move the geologist to the extremes—recommend a wildcat well, defer the prospect, or even drop it from the list.

Seismic

Seismic data, especially 3D data, is one of a handful of vital enablers that has made deepwater exploration so rewarding. No one ever *knows* what will turn up when the drillbit hits total depth, but seismic provides a quantum leap in improving probability of success before the well is drilled. With deepwater wells running up to $100 million, that's important.

Understanding seismic is more of a challenge than any other discipline in the oil and gas business. The professionals who work seismic—the geophysicists, geologists, petrophysicists, mathematicians, computer specialists, and other proselytized scientists—have their own language (Fig. 3–3). While there is little hope of totally demystifying seismic here, and at the risk of causing all of them to guffaw, a quick look at the mechanics and nomenclature follows.

Fig. 3–3 A Geologist and Geophysicist Labor Away at Their Subsurface Model at a Workstation (Courtesy Veritas DGC)

Seismic involves four steps: *acquisition, processing, display,* and *interpretation.* Companies that explore for hydrocarbons may outsource most of the first three to service companies that have climbed well up the seismic learning curve. Many also lean on those vendors to help them in interpretation.

Acquisition. Onshore or offshore, the object of this phase is to collect seismic data that present a picture of the subsurface. Seismic vessels like the one in Figure 3–4 tow *streamers,* plastic tubes that extend up to 30,000 ft in length behind the boat. The streamers float below the surface and have visible markers at the surface, placed intermittently to track the cables and to warn off vessels that might want to cross tracks and cut the cables.

Fig. 3–4 The Seismic Vessel *Seisquest* (Courtesy of Veritas DGC)

The plastic tubes contain hydrophones (pressure change detectors) with thousands in each streamer. The vessel also tows air guns, the seismic energy source, a short distance behind the boat. These guns are filled with compressed air, and when they open abruptly, they release the air with a bang, like a popped balloon. As that happens, the classic seismic data gathering takes place, as shown in Figure 3–5.

The sound wave travels down through the water, more or less unimpeded. As it passes the sea floor into the subsurface, the sound wave reflects or echoes off the boundaries between various layers of seabed, back to the surface where the hydrophones pick up the echo. Different hydrophones in a single streamer record the bounced seismic signal from different angles. The deeper the layer—and this is the essence—the longer the echo takes to reach a hydrophone.

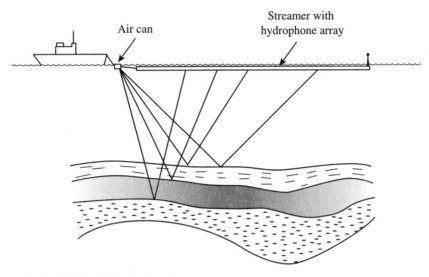

Fig. 3–5 Offshore Seismic Acquisition

Moreover, and this is critical as well, the various layers of the subsurface have different acoustical properties. The unique composition and density of each rock layer affects velocity of the echo and the time it takes to get to the hydrophone.

Individual *seismic records* appear as black wiggly lines. On the record in Figure 3–6, the intensity (or amplitude) of each line (geophysicists use the term *seismic trace*) measures the strength

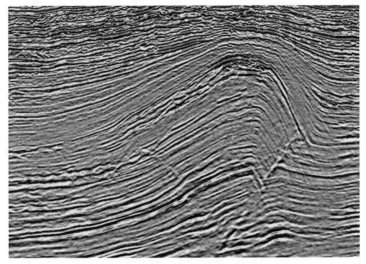

Fig. 3–6 A 2D Seismic Display of a Hanging Wall Anticline with Direct Hydrocarbon Indicator, West Niger Delta (Courtesy Veritas DGC)

of the reflected signal. The interval between the strong amplitude lines measures the time from one received signal to the next.

Multiple shots fire as the seismic vessel motors over a target area. Sometimes a vessel returns to a target and repeats the acquisition with a perpendicular pass. All in all, the number of recorded sounds can add up to millions of millions. The seismic travel time for a reflection at a depth of 10,000 ft is about three seconds. A deepwater seismic record typically has about 12 seconds worth of data.

Since the hydrophones string out behind the seismic vessel in a straight line, the seismic record comes in 2D format. The reflections come from below but not from the side. To create 3D records, seismic vessels string 10–15 parallel streamers behind them at 100- to 200-ft spacing. During processing, the parallel records are transformed into a continuous three-dimensional block of data, often called a seismic cube. Modern seismic cables convert the reflected sound waves into digital format using special modules in the cable. The digital data is transferred to the recording room on the boat. Some 3D surveys capture and store several terabytes of data. (Measuring information goes byte, kilobyte, megabyte, gigabyte, then terabyte, or 10^{12} pieces of information.)

Processing. Some processing takes place on the recording boat to organize the data, but most of it and all the subsequent reprocessing is performed using mainframe computers in the service company's or the operator's processing centers.

Initial processing is in the form of *data preparation* to eliminate bad records, correct for unwanted shallow surface effects, and reduce the effect of multiple reflections (wave fronts that bounce around within geologic layers and then back to the surface). This work is done using a series of software programs developed from mathematical theory about seismic wave fronts traveling through the earth.

After the initial clean-up, serious processing begins with a covey of tools meant to produce accurate representations of the subsurface.

• *Stacking.* One of the most important parts of processing is called *stacking* where as many as 100 recordings at varying offset distances are added to form one seismic trace. The primary purpose of stacking is to enhance the quality of the signal, reduce noise ratio, suppress multiple reflections that come from reflections within a layer, and in general, come up with a representation of the reflection sound waves from the surface to the subsurface and back to the surface.

• *Migration.* The stacking process assumes each reflection is at the midpoint of the traveled distance (down, then up) between the *detector* (the hydrophone) and the sound source (the air gun), after bouncing off the layer. That's fine if all the layers are flat. Reflections from *dipping beds* (layers that dip at an angle from the surface) can deliver a signal that can appear to be thousands of feet horizontally and hundreds of feet vertically away from reality.

To correct the seismic records and restore the signals to their true spatial relationships, processors use a scheme called *migration* (Fig. 3–7), an unfortunate term not to be confused with the movement of hydrocarbons from the source rock to the reservoir. To do migration (seismic-talk type), the processing programs have to use the velocity of sound in the various layers. Geophysicists and geologists sometimes have the luxury of data from a nearby well to estimate the velocities. Well logs have accurate measurement of depths and the compositions of various geologic layers. Using mathematical analysis, they can estimate the velocity of sound in each layer and approximate the depth/velocity relationship.

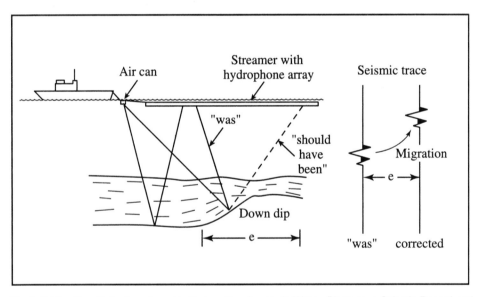

Fig. 3–7 Migration: Reflections from Updip Location Appear in Wrong Place on a Seismic Record and Must Be Corrected during Processing to Get Best Approximation of Where the Reflection Originated

In some deepwater areas, the conversion can be tricky. The Gulf of Mexico, for example, has many salt domes and salt layers that have both been deformed by the heavy weight of accumulated sands and shales deposited from rivers. Sound travels about 15,000 ft per second in salt but only 8000–12,000 ft per second in sediment. Seismic signals sometimes take a tortuous path through sediment and salt layers, and the seismic records can be very poor.

• *Prestack depth migration.* This deals with the distortions caused by inconsistent and odd-angled layers. The data is migrated (corrected) before stacking and is measured in depth. The corrections to the data require accurate subsurface velocity data and massive computer processing, an expensive way to adjust the data. A somewhat cheaper approach, *prestack time migration,* uses less rigorous velocity data and less computer processing.

• *Direct hydrocarbon indicators (DHI).* Geophysicists use DHI, which make it possible to predict hydrocarbon accumulations before a well is drilled. Bright spots are the most common DHI. The processing schemes take care to preserve the relative amplitudes of the data that make recognizing bright spots possible. The trained eye can locate likely oil and gas accumulations on a seismic display, as in Figure 3–6, although they do more detailed analysis and measurements before they make a prediction.

• *Amplitude versus offset.* Geophysicists also use a related technique, *amplitude versus offset* (AVO) to help directly identify hydrocarbons. They have learned that seismic signal amplitudes change with offset distance in a different manner for oil-, gas-, and water-bearing zones and these relationships help identify hydrocarbon-bearing zones.

• *4D seismic.* This adds the dimension of time to seismology. As a reservoir is evacuated by production, its acoustical image changes. Careful analysis of time-lapsed evacuation can identify pockets of oil not being drained and can improve future development plans. 4D seismic, multiple shoots of the same field, requires precise knowledge of the hydrophone locations as the seismic records are collected so that one seismic shoot can be compared to another.

Display. Results of the massive manipulation of seismic data must be displayed in a useful form. Processed 3D seismic data cubes are stored in the computer. Simple 2D vertical slices, like the one shown in Figure 3–6 provide first looks at the geology. Horizontal slices of the data cube, usually called time slices, can also be displayed on the computer screens. A 3D cube like that in Figure 3–8 can be rotated to get different views of the seismic image of the subsurface. Certain data characteristics such as strong seismic reflections can be isolated and displayed as a separate data cube using *voxel* technology. (A voxel is the 3D equivalent of the 2D pixel used to display images on home TV sets.)

Early on, interpreters found that the amount of data in black and white seismic lines overwhelmed the human eye. However, the eye can handle wide-ranging variations in color. Interpreters, even as late as the 1970s, used colored pencils to add another dimension to their displays. That sometimes led to disparaging, but good-natured remarks about the sophistication of seismic interpreters. Eventually software programs differentiated amplitudes, spacing, and other characteristics of the seismic reflections with color arrays chosen at the pleasure of the interpreter. (*See* Fig. 3–9.)

Now 3D seismic data along with geologic and engineering data are displayed in visualization rooms, like that in Figure 3–10. Images of the subsurface are shown on large high-fidelity screens and appear on one or more walls to create an illusion of the viewer being present in the subsurface looking at the data from different angles, sometimes using special 3D glasses.

Interpretation. All this preparation is just foreplay to the final step, interpretation and making economic decisions. Teams that can include geophysicists, geologists, petrophysicists, and other professionals bring their special knowledge to the interpretation as they search for the source rock, the reservoir, and the trap and direct indicators of hydrocarbon presence. That calls for intimate interaction to relate seismic data to the geologic and geophysical knowledge of the area. Experienced interpreters can recognize false signals given by bad processing and can order reprocessing with corrections.

In the end, seismic work plays a critical—and most often pivotal—role in the decision to drill a deepwater wildcat well or even a delineation or development well.

Fig. 3–8 3D Block of Seismic Data with Horizontal and
Vertical Slices

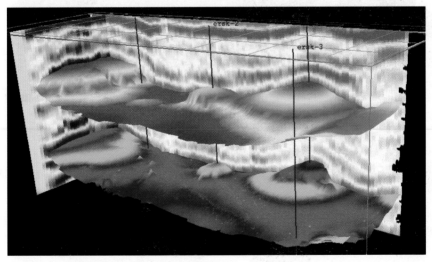

Fig. 3–9 Full Color Display of an Oil and Gas Field

Fig. 3–10 Seismic Display in a Visualization Room

Drilling a Wildcat

Suppose the geologist gets past these hurdles and convinces management to spend the money to drill a wildcat. The geologist and the geophysicist begin collaborating with a new set of scientists, engineers, and field operating people to plan and execute the wildcat well.

- The geologist and geophysicist bring the subsurface story to the table. They specify the depth, the geological objectives, i.e., what kind of formations they expect to drill through and the target depth they want to reach.

- They consult petrophysical engineers about downhole wellbore pressures, a program for core sampling, and electric log evaluation during the drilling sequence. That helps identify the presence of hydrocarbons and the quality of the target sands, i.e., the formations that might hold those hydrocarbons.

- Sometimes a paleontologist joins the team to identify fossils or remnants of animal or plant life in the core samples. That helps identify the age of the rock strata being drilled through, confirming or rejecting earlier assumptions or conclusions about the geological history and composition of the play. To the paleontologist, a wildcat provides an opportunity to collect geological information for use on subsequent wells drilled in the area.

- In contrast, the buck eventually gets passed to the field operating people, who bring their knowledge of the drilling vessels, drilling rigs, and techniques most suited for this prospect. In the end, they have to drill the wildcat and reach the target. More, much more on this subject follows in chapter 4.

Appraisal

The wildcat has been drilled. Hydrocarbons are present in the well test. The exploration geologist and geophysicist are thrilled with the success. But the excitement soon dissipates as management asks the critical questions, "How big is the accumulation? Are there enough reserves to justify further investment?"

As part of the planning process, the team members who worked the subsurface model developed various scenarios. If the well proves to contain hydrocarbons, where would they drill the next well, an appraisal well? Would more seismic be useful? Superimposed on these ideas, the team considers and integrates the information from the various logs from the wildcat into their model.

Based on these feedback and analyses, the team recommends one of several options:

• Plug and abandon the well because it has no further use in the appraisal process or in the development plans. Remarkably, this is very often the case with a wildcat well.

• Plug back the lower portion of the well and use the upper portion to drill a sidetrack to a new bottom hole location to help determine the size of the total accumulation.

• Temporarily abandon the well until further analysis indicates how it can best be used in the appraisal or development process.

• Complete the well. In some cases, if there is a high probability that field development will go forward, and if the well is situated in a good spot, the rig crew can leave the well prepared to be completed in the fashion explained in the following chapter.

An appraisal process can last for a period of a few months to more than three years, depending on the number of wells drilled and the amount of additional seismic data collected. The end game of the appraisal process is a decision to either abandon the prospect or proceed with a drilling program to develop the field, the subject of chapter 4.

Deepwater Plays in Context

The exploration process flow in Figure 3–1 and the description may seem straightforward and linear, and in microcosm, it is. But as most E& P companies contemplate the huge expenditures required to test a deepwater play, they choose to spread their risks—they pursue in parallel a number of exploration plays in various basins that have different risk profiles to avoid the consequences of "gambler's ruin," incurring a debilitating loss on a bet that had huge possibilities. A deepwater play exposes large sums of investment upfront, with low probability of success but huge potential rewards. Other plays, onshore or shallower water, might have higher probabilities of success but lower potential rewards. Exclusive commitment to a deepwater strategy could lead to financial ruin.

That was the story told in chapter 2 about Shell Oil Company, who pioneered the deepwater Gulf of Mexico in the 1970s and 1980s, but only in parallel with other domestic plays. In contrast, Petrobras had almost no alternatives to their low probability deepwater plays. At the same time, they faced a national mandate to find something and develop it. In hindsight, of all the plays pursued by all the E&P companies in the U.S. Lower 48 and in Brazil from the 1970s through the 1990s, the deepwater became the only exploration success.

Geology—the Shelf vs. the Deepwater

For 25 years, the oil and gas industry enjoyed the bountiful resources beneath the Gulf of Mexico. Success built on itself with the time-proven principle that an explorer can find no better place to look for hydrocarbons than in a basin where they have already been proven to exist.

Much of the richness of the Gulf of Mexico comes from its position as a passive margin—an area along the continent and ocean that does *not* coincide with the boundary between tectonic plates, those thick subsurface slabs that wander around the globe, albeit only about an inch or so per year. The Gulf of Mexico contrasts with the California coast, a geologically violent area where the Pacific plate and the North American plate continually grate on each other's edges. Basins like the Gulf of Mexico often have a profile like that shown in Figure 3–11, sloping more or less continuously from the onshore to the deepwater and beyond to an area called the abyssal zone, or even more ominously, the Abyss. The slopes, as far out as 12,000 ft in depth, provide depositories for thick layers of sediments, sand, shale, and silt from millions of years of onshore runoff. Beyond that, in the abyss, runoff must have been too diluted because deposits are thin and scarce.

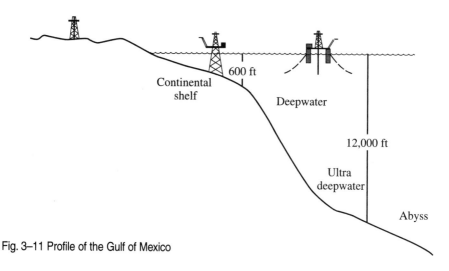

Fig. 3–11 Profile of the Gulf of Mexico

The Gulf of Mexico Continental Shelf, a more or less flat area, gently declines to a depth of about 600 ft where the continental slope begins. Oil companies focused for 25 years on the shelf before signs of maturity made them think about deeper prospects.

Characteristics of the geology of the area out to 600-ft depths and in the deepwater depths of 1200 ft and beyond—deltaic sands versus turbidite sands—differ in the fortuitous way described in chapter 2. The shallow part of the continental slope, between 600 and 1200 ft, called the *Flex Trend*, is a transition zone from deltaic to turbidite geology. More on turbidites later.

Salt, plain old sodium chloride, in massive accumulations, proliferates the Gulf of Mexico and provides distinctive similarities and differences between the shelf and the deepwater. Many geologists accept the view that during several interludes over the last 150 million years, the Gulf of Mexico became landlocked and dried up, leaving volumes of salt as the water evaporated. Over the last 25 million years, the onshore runoff of the ancient rivers deposited sediment on top of the salt.

After that, salt began its unique contribution to offshore geology. Salt has a strange reaction to heat and pressure, the type that results from deep layers of sediment above it. It has a propensity to move like the material in one of those psychedelic lava lamps that 50-year-old former hippies still have around. The salt accumulations slowly squirm upward, breaking through and deforming rocks, the sand, shale, and silt above. As the salt pushes up and pierces the overlying rock layers, it can form anticlines, salt domes, faulted areas (Fig. 3–12), and other structures delightfully capable of trapping hydrocarbons. The proliferation of salt on the shelf and in the deepwater provides one

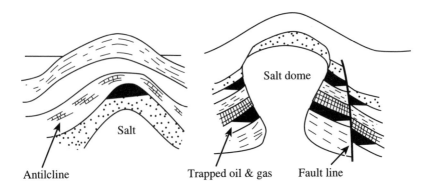

Fig. 3–12 Salt-Related Formations: an Anticline, a Dome, and Faulting

of the four crucial requirements the exploration geologist relentlessly searches for—the trapping mechanism.

The importance of salt in the Gulf of Mexico can hardly be overstated. The majority of oil and gas fields discovered on the shelf and the shallow depths of the Flex Trend have been in deltaic sand reservoirs on the flanks of salt domes. Deltaic sands have excellent porosity and permeability. Porosity measures the spaces between the sand that can hold liquids like oil or water. Permeability measures how well the spaces are interconnected. That determines how easily hydrocarbons flow through the sands.

The bad news about the movement of salt comes about by the faulting in Figure 3–12. Breaking up the sediment layers into small compartments or traps limits both the production rate and the ultimate recovery through a single well. With a small reservoir (compartment), high production rates risk early breakthrough to the wellbore of the gas that sits above the oil (*gas updip*) or the water that sits below (*water downdip*) as the gas or water bypasses some of the more sluggish oil. In either case, the oil will remain regrettably unrecoverable.

But more good news. The oil and gas fields in the Gulf of Mexico deepwater have been mostly found in turbidite sand reservoirs generally located in mini-basins between salt deposits. These turbidite sands reached the base of the continental slope in turbid currents—water heavily laden with sediments that increase its density and cause it to flow along the bottom of the sea. As the sediments deposited on the salt layers, they gently pushed out the salt, simultaneously creating large basins with minimal faulting, a critical aspect of the productivity of turbidite reservoirs. The sand quality, porosity and permeability of the turbidite reservoirs are similar to the deltaic sands, with generally good continuity. But because of the larger trap size, wells into these reservoirs can produce at much higher rates and achieve much larger ultimate recovery of the oil in place than their counterparts in the shelf.

The high production rates and ultimate recovery per well reduce the capital cost to drain the reservoir and the expense of operating a floating production platform in thousands of feet of water. That's what makes $25 million wells in the deepwater competitive with $2.5 million on the shelf and $0.5 million wells onshore.

And why does a deepwater well cost $25 million or more? That's another story, for the next chapter.

4

Drilling and Completing Wells

In completing one discovery

we never fail to get imperfect knowledge

of others of which

we could have no idea before.

From *Experiments and Observations on Different Kinds of Air*

Joseph Priestly 1733–1804

The exploration geologist and the team of subsurface professionals have completed their technical work in their preferred area; they have identified a high quality prospect; and they have recommended it to their management who has approved it, *in principle*, meaning everything is a go except the money. At this point, the exploration geologist pulls the trigger to start the process shown in Figure 4–1 by requesting a cost estimate and a well plan from another specialist, the drilling engineer. In many companies, the baton then passes to the drilling engineer who takes the

Key Process Steps	Responsible Party
Define the well objectives	Geologist/Geophysicist
Estimate the downhole pressures	Geologist/Geophysicist
Establish the well evaluation criteria	Geologist/Geophysicist
Specify the downhole logging program	Petrophysical Engineer
Identify any special issues	Everyone involved
Prepare the well plan	Drilling Engineer
Estimate the well cost	Drilling Engineer
Survey the drilling rig options	Drilling Engineer
Select the drilling rig	Drilling Engineer
Select service providers	Everyone involved
Drill the well	Drilling Operations
Evaluate the well	Drilling Operations
Complete or abandon the well	Drilling Operations

Fig. 4–1 The Drilling Process

lead in the project. The new team leader has the responsibility for getting the right people on the team to drill the well and for getting the attention of the right technical experts and ensuring they provide the appropriate technical support to the team. Often some members of the previous team carry over to the new team, especially the exploration geologist.

The Well Plan

Early on, the drilling engineer prepares the *drilling prognosis* for the well, a document of 20 or more pages that covers all the details and activities of the actual drilling phase. Figure 4–2 shows a typical cover sheet. The drilling prognosis lays out the requirements of all the contributors. It includes the specifics critical to the drilling plan:

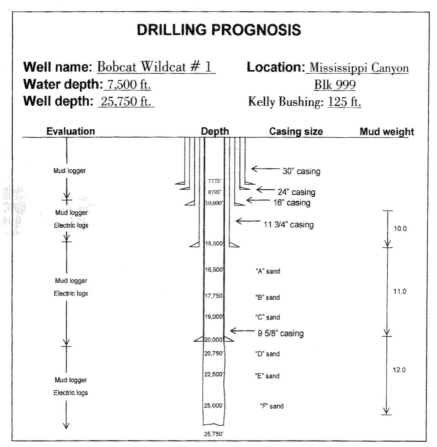

Fig. 4–2 Cover Sheet for a Drilling Prognosis

- The well location
- The water depth
- The vertical depth and the total measured depth of the well (Vertical depth is measured from the rig floor straight down to the target depth; total measured depth is the distance the drillbit travels to the bottom of the well—different from the vertical if the well is directionally drilled.)
- The depths of expected *reservoir* sands (there may be several)
- Down hole reservoir pressures and any unusual pressure zones or changes
- The expected hydrocarbons, oil or gas or both
- The presence of H_2S (hydrogen sulfide) or CO_2 (carbon dioxide) in the gas
- Evaluation needs (mud logs, electric logs, drillstem test, etc.)
- Special drilling problems such as loop currents, shallow hazards, or shallow water flows

The drilling prognosis also specifies the final disposition of the well. If the well finds hydrocarbons, is it kept as a producer or is it abandoned? Incredibly enough, abandonment could be the plan if it is drilled as an *expendable well*. In that case, the well is permanently plugged with cement and abandoned according to local government regulations.

The drilling prognosis includes any plans for future *sidetracking* of the well. In those areas where multiple reservoirs exist, drillers can sometimes access them by drilling below or through the side of an existing well casing. From there, they head in a different trajectory, having taken advantage of the borehole already drilled to that point.

The drilling engineer and the exploration geologist attach the drilling prognosis to their joint request for management to sign an authorization for expenditure (AFE) to drill the well.

Rig Selection

Rigs that can drill the deepwater have come a long way from the *CUSS I*, the first drillship, highlighted in chapter 1. The drilling engineer now has the choice of more than 150 semisubmersibles and drillships that can handle water depths of 1500 ft or more. Rigs like the semisubmersible *Nautilus* in Figure 4–3 and the drillship *Discoverer Spirit* in Figure 4–4 can drill wells down more than 30,000 ft (measured from the rig floor) in water 10,000 ft deep or more.

Fig. 4–3 The Semisubmersible *Nautilus* (Courtesy Transocean)

Fig. 4–4 The Drillship *Enterprise* (Courtesy Transocean)

To narrow the choices, the drilling engineer lists the project requirements the rig must meet:

- Water depths and well depths
- Pressure ratings, riser sizes, blowout preventer specifications
- Deck space and variable load capacity
- Drilling mud weight and delivery capacity
- Hook load capacity (how much pipe or casing the derrick can handle)
- Remote operated vehicle (ROV) capability
- Safety and environmental performance record
- Mobilization costs
- Day rate—the dollars per day to use it
- Length of contract
- Availability

With these criteria, the drilling engineer and drilling superintendent chose a group of companies with rigs that meet the demands and either get bids for the work or negotiate a short- or long–term contract for several wells or several years.

Once the potential suppliers submit their formal bids, the drilling superintendent and his team select the rig operator who they believe can drill the well in the most cost effective and safest way.

Many deepwater wells are drilled later, in the development phase, from the production platforms—a fixed platform, a compliant tower, a tension leg platform, or a spar. The drilling rigs used for those wells are given the generic name *platform rigs*. They have all the hydraulic, electrical, and load capacity of the floating rigs, but they come packaged to sit on the deck of a production system and not on their own exclusive hull.

Drilling

With the AFE in hand and the rig contract signed, the time to *spud the well* (begin drilling) arrives. If the rig contract calls for a semisubmersible, a large anchor-handling boat tows it to the location and assists with the mooring. Satellite-fed signals assist in the initial positioning of the semi and in the continuous monitoring during drilling operations. After all, as any deepwater driller will boast with the least bit of encouragement, drilling a well in 1500 ft of water is comparable to standing on top of the Sears Tower trying to stick a long straw in a bottle of Coke sitting on South Wacker Drive. Success in one case delivers the refreshing joy of a cola and in the other, whatever hydrocarbons riches a reservoir can deliver.

If the rig contract calls for a drillship, it moves to the drillsite under its own power and locates itself via dynamic positioning using its external thrusters on the fore, aft, and sides and the aid of continuous satellite tracking.

Semisubmersibles need to be moored on site, and some carry their own mooring rigging with them. Small anchor-handling boats assist in setting anchors around the drillsite to hold the rig in place. As semis grew in size to handle deeper waters, the kit required to moor them increased to mammoth proportions, demanding more deck space and flotation. In a forehead-slapping insight, rig companies began hiring separate work boats to haul the mooring apparatus ahead of time to the drillsite and set the anchors. When the semi arrived, hook-up took only hours, not days, and anchor boat day rates substituted for semisubmersible rates, saving about 80% of the mooring cost.

The crew that operates the drilling rig or drillship are employees of the drilling company that provides the rig. On board also are other employees of other service companies that run testing, drilling mud operations, or other special functions. The representation from the E&P company, ironically referred to as the *operator*, usually consists of one or two drilling foremen. The operator's (E&P company's) representatives have the final word on how the well is drilled, but the drilling company is expected to run a safe and efficient operation.

The petrophysical engineer, or another professional who plays that role, takes the lead in the well evaluation, both as the drilling proceeds and after the target depth is reached. In the drilling prognosis, the petrophysical engineer specifies intermittent or continuous electric logs and mud logs. (The drilling prognosis in Figure 4–2 shows the logging requirement on the left.)

The *measure while drilling* (MWD) tool assesses the lithology and determines the presence of hydrocarbons as the drillbit penetrates. Electric logs run at each casing point, and at the total depth, help assess the nature of the drilled-through layers, the presence of oil and gas in the reservoir sands, and the quality of the reservoir sands. Mud logs provide a record of the well cuttings and any hydrocarbon content as the well is being drilled. Usually service companies (Schlumberger, Halliburton, or dozens of smaller companies) perform these logging tests for the petrophysical engineer who then interprets the logs as part of the process to determine if the well is a success or failure.

Drilling Mud. Pressure control sits at the top of the list of worries for the drilling engineer. As the drillbit goes deeper, it encounters increasing pressure in the formation, due to the weight of the various rock layers and the column of water above it. Pressures increase more or less predictably in many areas, but in the deepwater, abnormal geopressures are often encountered.

As the well is drilled, drilling mud is circulated down the drillpipe and up the borehole *annulus* (the space between the drillpipe and the walls of the well). The weight of the mud is tuned to the pressures at the bottom of the hole. Three things can happen as the weight of the mud is varied. If the mud is *not heavy enough* to contain the pressures encountered, the wellbore may cave in, or worse, the oil and gas may come uncontrollably spewing up the wellbore, forcing the drillers to close the blowout preventer. If the mud is *too heavy*, it may overwhelm the strength of the rock, fracture the sides of the well, and leak off into the formation. If the mud is *just right*, the wellbore maintains its integrity, and any hydrocarbons encountered are kept in the formation until the well can be evaluated and completed.

As the mud flows down the drillpipe, out the jets on the drillbit, and up the borehole annulus, it performs two other functions. It cools the drillbit and carries away the drilling cuttings. The mud logger monitors the cuttings as they are separated from the mud at the surface. The cuttings help determine the nature of each layer drilled through, including the presence of hydrocarbons.

As the well goes deeper, heavier mud is needed to offset the higher pressures encountered. The weight of the mud has to be increased. But the mud is homogeneous and the heavier it is to accommodate pressures at the bottom of the hole, the more pressure it puts on the wellbore at intermediate depths. In fact, the weight of the mud may eventually increase to the point where it could fracture the rock uphole, at intermediate depths. To prevent that, steel casing is run in the well to preselected depths (after the drillpipe is pulled out of the wellbore), and cemented in place. The steel casing can readily handle the higher mud weight pressure and protects the weaker rock formations from fracturing.

The petrophysical engineer provides the drilling engineer with the *formation fracture gradient* information to determine when casing should be run. As the well proceeds, casing has to be run several times to cover weak formations and allow the drillbit to reach the targeted total depth. The geometry of the well forces each new string of casing to have a smaller diameter—it has to fit inside the previously run casing to get to the bottom of the borehole. The schematic on the drilling prognosis cover sheet in Figure 4–2 shows the general plan for running casing. In that sketch, the well starts with 30-in. *surface casing*. By the time the well reaches 20,000 ft, four more sets of casing, each of decreasing size are put in place. Setting casing for the final 5750 ft of wellbore depends on whether the well encounters hydrocarbons and is to be completed

Blowout preventers. As another precaution against unusual and even catastrophic surges in wellbore pressure, every well is fitted with a *blowout preventer* (BOP) system. The BOP can seal off

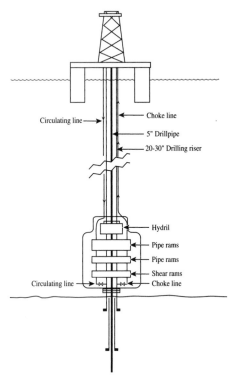

Fig. 4–5 Blowout Preventer System

fluid flow from the well through one or more devices activated from the drilling rig control room. The choice of which one to use depends on the severity of the situation, but all are present to protect the drilling operations, personnel, and the environment.

For a well being drilled from a semisubmersible or a drillship, the BOP is placed on the well casing head at the sea floor. For a platform rig on a stable drilling platform like a fixed-to-bottom platform, a tension leg platform, or a spar, the BOP can be located at the surface, just below the drilling rig.

The *BOP stack* has three or more sets of hydraulic devices, as shown in Figure 4–5. The first line of defense is the *Hydril* or *annular preventer*. On activation, this donut-shaped rubber-steel composite closes off the annular space between the drillpipe and the borehole. The Hydril is closed by the driller as soon as he sees any sign of a surge in the drilling mud flow. That indicates a higher pressure zone has been penetrated and mud weight needs to be increased.

If the pressures exceed what the Hydril can control or if the Hydril has been damaged, the next safeguards are the pipe *rams*, made of steel and rubber and contoured to fit around the drillpipe and seal off the annulus. With the Hydril or the pipe ram closed, the valve on the *choke line* is opened. That allows heavier weight drilling mud to be circulated down the drillpipe and back up the section of the annulus below the BOP stack, and then up the choke line to the surface. Once the heavier weight mud controls the higher reservoir pressures, the pipe rams and Hydril can be reopened safely.

In extraordinary circumstances, *shear rams* provide the last resort. These steel blinds cut through the walls of the steel drillpipe and seal off both the annulus and the drillpipe. Using the shear rams causes irreversible damage to the drilling sequence. The well will have to be re-entered; the pressure will have to be overcome; the severed drillpipe will have to be removed. All this can add costly days to the drilling operation.

Drilling crews test the BOP system on rigorous schedules to ensure it always functions. Company policies and local government agencies determine the frequency and testing procedures.

Evaluating the well

After the drillbit reaches the target depth (TD), the bit is pulled and the drilling engineer and her team evaluate the well. A drillstem test may evaluate the flow rates of hydrocarbons from the zones not yet covered by casing. Integrating this data with the logs and other tests leads to the completion decision. The drilling vessel can stand by while this decision is being mulled, or the well can be "temporarily abandoned" by placing cement plugs in the wellbore, and then disconnecting at the BOP, with the vessel moving on to another assignment.

Completing the Well

Another decision point arrives for the subsurface and drilling teams—whether or not to complete the well. Depending on the circumstances, well completion can start as soon as the AFE can be signed. In the case of a wildcat, creating a complete development plan may delay completion of the well for two or three years.

What's left to *complete?* To produce oil and gas effectively, a well has to have additional casing run; tubing, through which the production flows, has to be put in place; the casing has to be perforated below the tubing so the oil and gas can flow; a *tree* has to be installed at the top of the well, safety devices need to be put in place; and at the reservoir sands to be completed, a kit has to be installed to keep sand from clogging up the well. The process to carry out these few steps, shown in Figure 4–6, differs from onshore and shallow water completions primarily in the complexity and cost.

Even before the drilling phase finishes, the geologist and reservoir engineer have raw data to help them work on the first

---

Drillstem Test

"A procedure to determine the productive capacity, pressure, permeability, or extent (or a combination of these) of a hydrocarbons reservoir. While several different proprietary hardware sets are available to accomplish this, the common idea is to isolate the zone of interest with temporary packers. Next, one or more valves are opened to produce the reservoir fluids through the drillpipe and allow the well to flow for a time. Finally, the operator kills the well, closes the valves, removes the packers and trips the tools out of the hole. Depending on the requirements and goals for the test, it may be of short (one hour or less) or long (several days or weeks) duration, and there might be more than one flow period and pressure buildup period."

(From Schlumberger *Glossary of Oilfield Terms* at www.glossary.oilfield.slb.com/)

Process Steps	Responsible Party
Create a reservoir model	Geologist or Reservoir Engineer
Specify reservoir pressures and temperatures	Reservoir Engineer
Estimate well rates and ultimate recovery	Reservoir Engineer
Identify any special issues	All involved
Develop a completion plan	Completion Engineer
Estimate the completion cost	Completion Engineer
Approve the drilling company capabilities	Completion Engineer
Select service providers	All involved
Complete the well	Completion Operations

Fig. 4–6 Well Completion Process

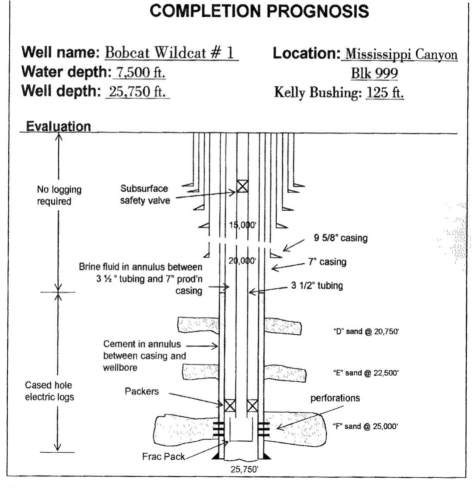

COMPLETION PROGNOSIS

Well name: Bobcat Wildcat # 1   **Location:** Mississippi Canyon
Water depth: 7,500 ft.   Blk 999
Well depth: 25,750 ft.   Kelly Bushing: 125 ft.

Evaluation

No logging required

Subsurface safety valve

15,000'

9 5/8" casing

Brine fluid in annulus between 3 ½ " tubing and 7" prod'n casing

20,000'

7" casing

3 1/2" tubing

"D" sand @ 20,750'

Cement in annulus between casing and wellbore

"E" sand @ 22,500'

Cased hole electric logs

Packers

perforations

"F" sand @ 25,000'

Frac Pack

25,750'

Fig. 4–7 Cover Sheet for a Well Completion Prognosis

three steps in the well completion process, defining how the reservoir behaves. Their information, plus input from the petrophysical engineer, the geophysicist, and others become the fodder for a completion plan.

Once again, the baton passes in most companies to a completion (or production) engineer, usually a mechanically inclined person. The first chore of the completion engineer is the preparation of a *completion prognosis*. Figure 4–7 shows the cover sheet. Like the drilling prognosis, this document lays out the details of the work plan.

- Who is involved and what their responsibilities are
- Procedures for completion
- The mechanicals
    - Completion fluid type and weight
    - Tubing and casing specifications
        - Perforation type and depth
        - Tubing size
        - Packer type and locations
        - Subsurface safety valve type and depth
    - Tree specifications
- Procedures for start-up

The well completion prognosis becomes the basis for preparing the AFE for well completion funds.

The mechanicals

The well has to be emptied of drilling mud before it can be completed. The completion fluid used to displace it is a brine composed of water and either sodium chloride, calcium chloride, or zinc bromide. During later steps, the possibility of leaking some of this fluid into the subsurface formations warrants the environmental-friendly characteristics of these mixtures and for a fluid that does not damage the reservoirs. The concentration of these completion brines can be increased to achieve the same weight as the mud being removed. The brine performs the same pressure control function as drilling mud while still allowing other completion work to be done in the wellbore.

During the drilling phase, protective casing was run down the hole, probably about 80% of the way. The first operational item is running production casing inside the existing casing, down below the deepest reservoir that contains hydrocarbons. This casing is the smallest in diameter, from 5 to 9 1/2 in., depending on the casing already installed above it. (*See* Fig. 4–7.)

The sequence of tasks that follows includes:

- cement the production casing in place
- run a tubing workstring
- displace the drilling mud with brine through the workstring
- perforate the casing
- set the *gravel pack* or the *frac pack*
- replace the workstring with production tubing and a packer
- displace the completion fluid with a permanent, corrosion-resistant fluid
- remove the BOP
- set the tree

Christmas Trees—Wet Trees—Dry Trees

In the beginning, the device at the surface of a well consisted of a central stem plus a few valves, nozzles, and handles, often wider at the bottom and sometimes resembling a tree. Oil patch legend has it that in Oklahoma and Texas, oilfield hands festooned them with lights and garlands during the Yuletide season, begging for the enduring name, *Christmas tree*, and later just *tree*.

People seldom mistake today's surface devices for a forest product, and the self-contained apparatus used on subsea wells has no resemblance at all. In the offshore, *dry trees* sit on a production platform where the crew can use their hands to work them. *Wet trees* sit on the wellhead at the sea floor. An operator works them from a production platform connected through an *umbilical*. Chapter 8 describes the system in more detail.

As one of many safety devices a few thousand feet below the seabed, a *subsurface safety valve* is located in the wellbore tubing string. (*See* Fig. 4–7.) This failsafe valve closes automatically when the pressure holding it open drops precipitously. That stops the flow of oil and gas from the well if a catastrophic event occurs such as the tree or the platform being destroyed.

The completion engineer has to specify in the completion prognosis a *downhole completion* system. Its purpose is to prevent sand from the reservoir from flowing into the wellbore with the hydrocarbons. Geologists usually refer to reservoirs as rock, but in fact many are unconsolidated sand. Most have high geopressures waiting to be released. The prolific turbidites in the deepwater allow fluids to gush into the wellbore at rates of 5000–30,000 (or even more) barrels per day. At those rates, sand can easily fill up the wellbore or act like a sandblaster, eating away the innards of the well. In either of these disastrous circumstances, the well stops producing and the operator faces the choice of abandonment or a multimillion dollar recompletion.

The challenge for the completion engineer is not only to avoid or stop the production of sand but to also maintain a high well productivity. Shutting off sand flow is worthless if it kills the oil flow. A number of methods are available to control sand production in wells, but the three most commonly used are the conventional gravel pack, the frac pack, and wire-packed screens.

In gravel pack operations, shown in Figure 4–8, a steel screen is lowered into the well and the surrounding annular space between the screen, and the production casing is packed with gravel and sand of a specific size and gradation,

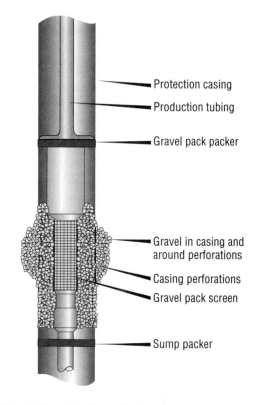

Fig. 4–8 Downhole Gravel Pack

selected to prevent the influx of reservoir sand and silt. The mesh in the screen is sized to retain the gravel and sand pack, yet also minimize any restriction to production.

In a frac pack installation, specially formulated slurry containing selected sand and gravel is pumped under high pressure and rate into a well's reservoir interval. That causes a vertical fracture to occur, splitting the reservoir open on both sides of the well. The sand in the slurry fills up the fractures. When the pressure is removed and the slurry flows out, the sand remains to "prop" open the fracture. The process creates a plane of high permeability in this fissure, an avenue from the reservoir through which the well fluids can flow. A frac pack also includes a downhole gravel pack to filter out the sand.

In wire-packed screen applications, wire mesh holding gravel in place is wrapped around a perforated liner for several turns and then welded to it.

As nearly the last step in the completion phase, the blowout preventer has to be removed and replaced with a tree, the device that controls the flow of the well. To ensure the integrity of the well, BOP removal and replacement with a tree has to be done at the same time.

With all the equipment in place, the completion fluid is removed by pumping water treated to prevent corrosion (or sometimes diesel fuel) down the production tubing annalus, displacing the brine through the tubing. A packer is set near the bottom of the production tubing (but above the perforations and sand control apparatus), sealing off the annulus. The treated fluid remains in the annulus permanently. As the pressure at the top of the production tubing is reduced, hydrocarbons push the treated brine out the production tubing as the well comes on stream. The completion engineer has reached the finish line, and he hands the baton to the operations personnel.

Special Problems

As E&P companies moved into deeper water, they encountered geologic and environmental obstacles never seen before. Some still vex them.

Loop current and eddies. A clockwise current of water from the Caribbean Sea into the Gulf of Mexico between Cuba and the Yucatan creates a flow known to ocean scientists and meteorologists as the Loop Current. Panel 1 of Figure 4–9 shows two stages of the Loop Current as it extends further into the Gulf. The movement of this current is more or less random both in timing and the extent to which it penetrates the Gulf. On its exit from the Gulf, it changes name to the Florida Current, a tributary of the Gulf Stream.

As in Panel 2, *eddy currents* spin off the Loop Current and work their way westward as a huge column of warm water spinning in a clockwise fashion. In Panel 3, the larger eddy current creates smaller, cold-water eddies, rotating in a counterclockwise direction. Eventually, both the cold and warm water eddies move westward into shallow water (Panel 4) and dissipate.

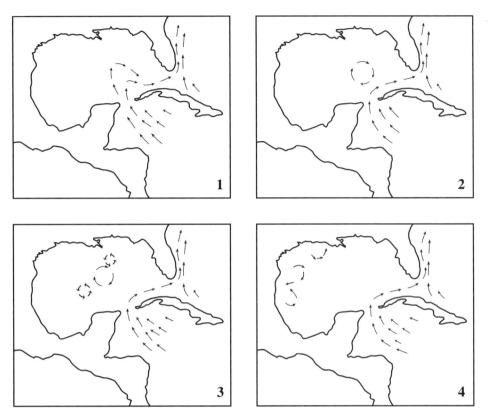

Fig. 4–9 Loop Current and Eddy Currents in the Gulf of Mexico

The velocity of the current as the eddy spins can be 2 to 4 knots. As the eddies pass by an oil and gas drilling or producing operation, they subject the facilities to unusual stress and vibration. In addition, depending on the position of the platform and the eddy, the current may increase in one direction, and then perhaps change to the opposite some time later.

The swirling eddy currents drift slowly westward in the Gulf at about one knot. Their diameters vary. As a result, a platform or drilling operation may see the effects of one eddy current for a day and the next one for a week or a month.

During an eddy current episode, the drilling riser from a semisubmersible or a drillship may bend or bow from the current to such an extent that the vessel has to change positions to stay connected. In some cases, the distortion can be so exaggerated that the drillpipe rubs against the drilling riser, a situation that warrants shutdown of operations before failure. At a producing operation, an eddy can cause a riser to vibrate, inducing worries about metal fatigue and ultimate failure. Some mechanical devices can deflect the vibration effect of the currents on risers but add to the drag.

The random nature of the eddy currents vexes the industry and begs for mechanical solutions to mitigate its effects. No known method to prevent the eddy currents themselves is on the horizon or even likely. E&P people just learn to live with them.

Shallow hazards. Occasionally drillers run into drillsites that have special geologic problems at depths of less than 2000 ft below the sea floor—excessive faulting makes drilling and well control more difficult; thin layers of gas can disrupt what should be easy drilling.

Special processing of seismic data helps identify the presence of these hazards, and when they are found, the solution is often to spud the well in an offset location and directionally drill the well around them to the target bottom hole location. Eliminating the delays that these shallow hazards sometimes cause usually offsets the extra cost of the directionally drilled well.

Shallow water flows. Another geologic quirk happens when sand layers in the first 2500 ft become slightly over-pressured during their original deposition. As the drillbit penetrates these layers, the contained water wants to flow into the wellbore. Increasing the weight of the drilling mud is the normal antidote to prevent inflows, but often the rock and sands at these depths are very young geologically and therefore have little strength. The extra mud weight can fracture other layers in the vicinity of the shallow water flows, causing loss of drilling mud and other well control problems. The situation calls for carefully setting casing at selected depths to isolate the troublesome sands, a time-consuming and costly solution, but a necessary procedure if the well is to reach the TD.

5

Development Systems

The time has come, the Walrus said

To speak of many things:

Of shoes – and ships – and sealing wax –

Of cabbages – and kings –

and why the sea is boiling hot –

And whether pigs have wings.

From *Through the Looking Glass,*

Lewis Carroll (1832–1898)

The exploration team has found a field, appraised it to the satisfaction of its management, and recommends development. In some companies, the explorers hand off this work to a development team in the production organization. In others, some members of the original

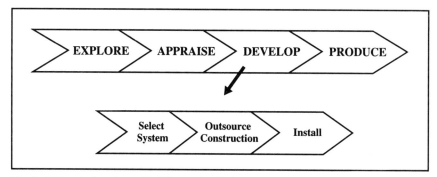

Fig. 5–1 Steps in Development of Deepwater Projects

exploration team follow the work and join the development team, in which case, a subsurface reservoir engineer or construction engineer leads this step. Whichever, the new team begins work on the first of three steps in Figure 5–1, selection of the development system.

Of course, before they ever spent money on the appraisal step, the geologists and engineers of the exploration team had to give some thought to how they would produce the field. Otherwise how would they estimate the ultimate profitability of the venture? While Figure 5–1 shows a linear sequence, reality is a bit messier than that.

Still, when the appraisal effort indicates a viable opportunity, work on the first step of development—selecting a system—escalates.

Development System Choices

The options available to develop and produce oil and gas in water a thousand feet or more deep fall into three broad groups, fixed to the sea floor, moored and tethered floating systems, and subsea systems, as shown in Figure 5–2.

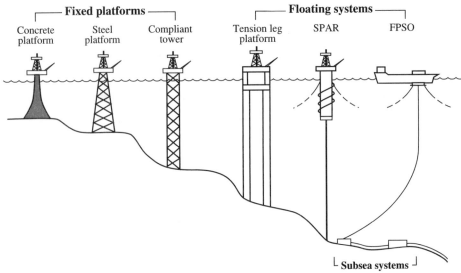

Fig. 5–2 Development System Options for Deepwater Projects

Fixed to the sea floor

These structures physically sit on the bottom. They are held in place either by the sheer weight of the structure or by steel piles driven into the seabed and affixed to the structure. The group includes the following.

Fixed platforms. These consist of a jacket and a deck. The jacket is the tall, vertical section built from tubular steel members and is locked to the seabed by driven piles. The topsides or deck placed on top has production equipment, crew quarters, and drilling rigs.

Compliant towers. Like the fixed platform, these are made of tubular steel members and are fixed to the bottom with piling, and support a deck. By design, compliant towers sustain more lateral deflection than the more massive and rigid fixed platforms.

Gravity platforms. These are built from reinforced concrete. With their resulting weight, gravity platforms rely on gravity to hold them in place. Gravity platforms are used in up to 1000 ft of water, but the seabed has to be especially firm to ensure no creep over time.

Floating systems

These include TLPs, FPS's, spars, and FPSOs. All four have to be moored in place with tendons or wire rope and chains in order to stay connected to the wells below.

Tension leg platforms (TLPs). These have floating hulls made of buoyant columns and pontoons. Steel pipe tendons hold the hulls below their natural level of flotation, keeping the tendons in tension and the hulls in place. Even so, the platforms experience some lateral motion in heavy seas. TLPs most often have dry trees on the platform, but wet tree subsea tie backs are also common. Like the fixed systems, TLPs can accommodate drilling operations from the deck. A variation of the TLP design, the smaller mini-TLP accommodates smaller deepwater reserves.

Spar platforms. These get their flotation from large diameter cylinders, weighted at the bottom to keep them upright. Eight to 16 wire or synthetic rope and chain combinations moor the hulls to the seabed. Because of their huge masses, spars have very little vertical heave, even in heavy seas, but there can be lateral offsets due to winds and currents. Like TLPs, dry trees predominate, but wet trees are also used, especially in deeper waters.

Floating production systems (FPS's). These consist of ship shape, TLP-like, or a semisubmersible hulls with production facilities on board. Wire or synthetic rope and chain moor them in place. FPS's are quite free to move both laterally and vertically, so only wet trees can be accommodated. No drilling either. The significant motion during heavy seas and currents calls for special equipment to accommodate the risers that get the oil and gas from the sea floor wet trees to the production facilities on the deck.

Floating platform, storage, and offloading (FPSOs). These have large ship shapes, made from either converted tankers or new construction. They are moored with rope and chain. Like FPS's, FPSOs have no drilling capability. They process production from subsea wells and store large crude oil volumes, accumulated for later transport by shuttle tankers. A variation, FSOs, receive processed oil from nearby platforms, often FPS's, and store it for subsequent transport by shuttle tankers. An FPS and FSO together are the equivalent of an FPSO. Water depths present no limitation to FPSOs and FSOs.

Subsea systems

This option can have single or multiple wellheads on the sea floor connected directly to a host platform or to a subsea manifold. The systems include connections by flowlines and risers to fixed or floating systems that could be miles away. Subsea systems can be set in any depth water.

More complete descriptions of each of these options follows in the next three chapters. This chapter sets those chapters up by looking at the criteria for selecting among them.

Choosing Development Systems

Selecting the right development systems involves assessment of the usual list of physical circumstances—water depth, reservoir configuration and location, access to oil transportation and, separately, gas transportation—plus the constraints placed by the local government and the institutional preferences of the investing operator.

Companies may each arrive at different conclusions for the same circumstances, reflecting their own predilections or biases. For example, Petrobras has always preferred wet tree subsea systems, while many Gulf of Mexico producers prefer dry trees on TLPs or spars. In some areas of the world, producers would rather not bring oil ashore so they choose floating production systems that offload into tankers. Regulatory agencies have to be convinced as well. For years the U.S. Minerals Management Service was hard pressed to allow the floating production systems that offload into tankers in the Gulf of Mexico; they preferred floating systems tied to pipelines. At the same time, the U.K. government had no problem with floating production systems and shuttle tankers in the North Sea.

Water depths. The realities of the laws of physics and the limited strength and flexibility of materials make water depths the first cut in choosing a development system. In round numbers, the maximum water depths for each of the options fall as follows:

> Fixed platforms..............................up to about 1500 ft
> Gravity platforms..............................1000 ft
> CTs..............................3000 ft
> TLPs..............................5000 ft
> Spars..............................7500 ft
> FPSs..............................Unlimited
> Subsea systems..............................Unlimited

At first glance, the water depth seems to be related to the amount of materials necessary to build the system. That may be true, but beyond that, the other maritime conditions—wind, waves, currents, and the competency of the sea bed to support heavy loads—figure importantly. Winter storms in the North Sea and the Atlantic off Nova Scotia sometimes generate winds of 125 mph, and those create waves 90 ft high. Hurricanes in the Gulf of Mexico can deliver 150 mph winds and 80 ft waves. In the more serene areas off the coast of West Africa, winds seldom exceed 75 mph and waves only reach 25 ft.

Thinking even more deeply, the extremely hard clay soil of the North Sea bottom provides fine support for gravity based structures. In contrast, the under-consolidated, soupy clay soils in the Gulf of Mexico would have platforms slipping and sliding around if they weren't nailed down with deep driven piles.

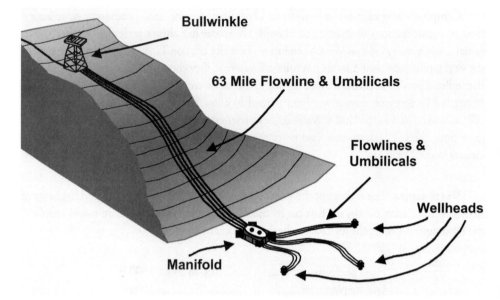

Fig. 5–3 The *Mensa* Field, a Subsea Development (Courtesy Shell Exploration and Production Company)

Oil transportation. The next cut at choosing development systems comes from the oil transportation options available. The Gulf of Mexico and the North Sea already have a web of pipelines in place. For many new discoveries, the option of tying into one of these lines might look attractive, even if miles of new subsea pipe have to be laid.

Sometimes the prohibitive cost of laying a new pipeline calls for floating production systems with oil storage onboard or floating nearby. In areas as remote as West Africa, West of the Shetlands, or the Faroe Islands, the complete lack of oil pipeline infrastructure calls for FPSOs, or at minimum, FSOs.

Gas disposition. Deepwater gas can provide good news or bad news in the selection of development systems. First of all, if the prospect has associated gas dissolved in the oil, something besides flaring has to be done with the gas as it is produced. Almost every offshore jurisdiction prohibits or restricts continuous flaring. Gas is good news when a gas pipeline infrastructure presents itself in some proximity. The good news is then compounded when the reservoir is predominantly gas, with just some associated liquids.

Gas flows in a rather frictionless way through miles of flowline with minimal loss of pressure. For example, the designers of the *Mensa* development, a gas reservoir in 5400 ft of Gulf of Mexico water, shrewdly avoided the expense of any onsite platform. They did a subsea development and tied it via flowline to a host platform 70 miles away, powered just by the reservoir's pressure with no boosting. (*See* Fig. 5–3.) The receiving station happens to be a fixed platform in 1354 ft of water already pumping gas to the onshore via pipeline. In contrast, oil might flow under those conditions only 15–20 miles.

But what if no pipeline infrastructure for the gas exists in the region? In that case, if it is a gas-only well, nothing happens. The E&P company hangs a sign on it that says "Stranded Gas" and waits for the market to develop or for the technology to come forward that makes the gas economically transportable.

If the gas is associated with oil, and if the oil alone presents itself as an economically viable project, the gas can be reinjected (carefully) into the reservoir from whence it came or into another convenient reservoir. There it sits indefinitely, like the rest of the stranded gas in the world. Reinjection facilities on the producing facility consist primarily of compressors to force the gas back into the reservoirs.

In some cases, the volume of associated gas is small enough that the power requirements on the platform—generating electricity for pumps, compressors, heat and light—consume all of it.

Reservoir proximity. Often development opportunities can include more than one reservoir. In that case, the proximity of the various reservoirs can dictate the scheme selected. If they are relatively close to each other (say within two miles), directional drilling from one location makes sense. Each well can then be produced through a dry tree on a platform, making production operations less complicated. That arrangement favors a fixed platform, a TLP, a spar, or a compliant tower, but doesn't exclude development with subsea wet trees. If the reservoirs range over many miles, subsea wells connected via flowlines to a fixed or floating station makes more sense. As exploration in an area proceeds, various combinations of both wet and dry trees often emerge.

Development Examples

1. An oil and associated gas field located in 3000 ft of water 20 miles from existing shallow water platform with pipelines to the onshore.

   Probable solutions:
   - A TLP, a spar, or a compliant tower with dry trees; separate oil and gas pipelines to the nearby platform

   Possible alternatives:
   - Floating production system with wet trees; separate oil and gas export pipelines to the nearby platform
   - Subsea development (wet trees) with a single flowline back to the nearby platform

   Criteria:
   - Costs
   - Technical and commercial risk assessment
   - Timing
   - Experience and learning curve
   - Producer's bias toward various systems

2. An oilfield with associated gas located in 8000 ft of water 60 miles from any other infrastructure, and further from any market.

   Probable solution:
   - A floating production system (FSO or FPSO) with subsea trees, shuttle tankers for the oil, and reinjection of the gas

   Possible alternatives:
   - None

   Criteria:
   - Depth
   - The high cost of building pipelines eliminates other alternatives
   - No viable transport options for the gas

With this crystal clear selection criteria now laid out, the next two chapters examine the devil in the detail, taking a close look at the design of fixed structures, and floating and subsea systems.

6

Fixed Structures

We were the first that ever burst into that sea.

From *The Ancient Mariner*

Samuel Taylor Coleridge (1772–1834)

From the get-go, offshore operations employed fixed structures, secured to the seabed by pilings. Even exploratory wells were drilled from fixed platforms in the early days. Today E&P companies use fixed structures up to the threshold of the deepwater for both drilling and subsequent production of oil and gas.

The Conventional Platform

The fixed structure installed most often around the world consists of the following, as shown in Figure 6–1:

- The *jacket*, the steel structure that rises from the seabed to above the water line
- The *deck*, where the drilling and production equipment sits, located atop the jacket
- The *pilings*, steel cylinders that secure the platform to the seabed
- The *conductors* or *risers*, steel pipes through which the wells are drilled, completed, and produced

In contrast to onshore structures, the jacket is built from tubular steel members, not structural shapes. These cylindrical shapes create the lowest resistance to waves and currents, reducing the amount of steel and ultimate weight and bill for the platform. The deck legs, which connect to the jacket, are typically tubular, to allow easier geometric fit. Besides that, in many cases the deck

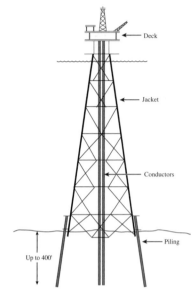

Fig. 6–1 Fixed Steel Platform

legs extend downward to near the waterline and are vulnerable to wave action in high seas. The deck itself is above water, so traditional square-edged structural pieces are as common as those found on an onshore operation or any heavily loaded high-rise building.

These steel jackets are typically built on their side, that is, rotated 90 degrees from their final installed position. This keeps the construction yard heights under control, but it does require an innovative approach to load out and launch these structures.

The Concrete Platform

In the 1970s, another form of fixed structure started to dot the seascape of the North Sea, steel reinforced concrete platforms. (*See* Fig. 6–2.) The unusually hard clay seabed of much of the North Sea allowed these heavy structures to plop down and stay in place by their sheer weight, despite the large profile they presented to waves and currents. Few have any pilings driven into the seabed.

Both these fixed structures, steel and concrete, rely on brute force and huge footprints to resist wind, wave, and current forces. For example, the *Bullwinkle* platform in 1354 ft of water in the Gulf of Mexico has a base of 400 ft by 480 ft. Twenty-eight steel pilings hold the jacket in place. Each cylindrical piling is 7 ft in diameter, with 2-in. thick walls, and is driven 400 ft into the sea floor. The *Troll* platform in the North Sea, a concrete gravity structure, has a base

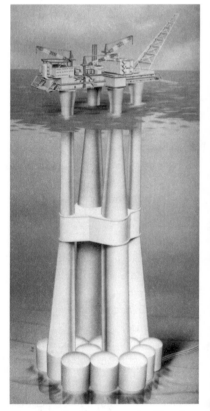

Fig. 6–2 Concrete Gravity Structure
(Courtesy Norske Shell)

approximately 500 ft in diameter. When it landed on the seabed, it penetrated tens of feet into the bottom. (*See* Fig. 6–2.)

The Compliant Tower

As oil and gas development moved into waters beyond 1500 ft deep, steel platforms and certainly concrete platforms required more materials and cost than the ventures could economically bear. The problem wasn't engineering. The base of the structures just became too large. Along came the idea of compliant towers, tall structures built of cylindrical steel members, but slender in shape. Pilings tie it to the seabed, but in a small footprint. A typical compliant tower in 1650 ft of water in the Gulf of Mexico water covers only 140 by 140 ft at its base. (*See* Fig. 6–3.) With that narrow a base, the compliant tower has none of the brute strength of the steel platform or concrete structure. In fact it sways with the currents, waves, and winds, as much as 10–15 ft off center in extreme cases, but during normal operating conditions the motions are very slight.

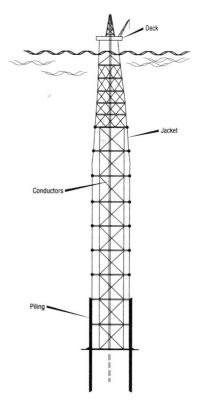

Compliant towers (CT) are designed to have considerable "mass" and buoyancy in their upper regions. The net result is that they have a very sluggish response to any forcing function. The typical 10- to 15-second cycle waves pass through the structural frame before it can respond, something like a water reed in a wave environment. (In fact, an early version of the CT was named the *Roseau*, the French word for reed.)

While fixed platforms have a technical and commercial limit of about 1500 ft of water—it just takes too much steel to go deeper—compliant towers can be used to about 3000 ft before the increasingly thicker steel members at its base make the concept infeasible. Ironically, compliant towers have a lower water limit of about 1000 ft. In shallower water, the flexibility needed for deeper water disappears. It becomes too stiff to handle the waves and currents and the concept doesn't work.

Fig. 6–3 Compliant Tower

Construction

Building steel platforms and CTs requires tedious fabrication of the cylindrical members. Steelworkers take flat steel plates and form them into cylindrical shapes of various diameters and lengths of 5–15 ft. Next, they *stalk* (weld together) these individual cans into tubular members. Then they cope (cut to shape) each end to fit up against other cylindrical members and form the prescribed and complex geometry of the structure. Coping becomes crucial. In the next step, they weld these intricate shapes to form the panels or sides of the structure. Typically, four panels make up the structure.

Once the fabricated panels sit completed on the ground, construction cranes roll them up for connection with each other by welding and addition of more cylindrical steel members. (*See* Fig. 6–4.) Tall as it is when upright, the structure typically lies on its side in the construction yard to accommodate construction access and, more importantly, transit to the offshore wellsite. Sites for the construction and assembly of these platforms have to be on waterways that allow access to the sea. More on the transport of these structures—ungainly when they are out of water—later.

Fig. 6–4 Construction Cranes "Rolling up" a Steel Frame for a Deepwater Jacket (Courtesy Shell Exploration and Production Company)

In round numbers, a steel platform in 300 ft of water requires about 3000 tons of steel for the jacket structure and another 1000 tons of steel pilings. For a steel platform in 1500 ft of water, the jacket structure increases to about 50,000 tons of steel and 15,000 tons of pilings. For comparison, the Eiffel Tower is only 1063 feet tall and has a mere 7300 tons of steel.

The steel jacket legs for a 300-ft water depth structure are about 54 in. in diameter, with the truss work being made up of 24- to 36-in. cylindrical braces. For a 1500-ft water depth case, the legs are 80–100 in. in diameter, the braces 30 to 60 in., and the piling about 70 or 80 in. Big stuff. A CT has individual members of the same dimensions as a fixed platform, but does not take as much overall steel because of its slenderness. A CT in 1700 ft of water requires about 30,000 tons of structural steel and 7000 tons of piling, considerably less than the 1500-ft fixed platform.

Gravity platforms. Using concrete for gravity structures came about because of a happy confluence of marine topography and geologic history in the North Sea. The seabed above reservoirs like Brent and Ekofisk consists of firm, hard, clay. Not far away, on the eastern side of the North Sea, deep fjords cut into the mountainous coastline.

Concrete structures have to be built like concrete buildings, from the bottom up, using reinforcing steel embedded in the concrete. While buildings rise toward the sky as they form, concrete platforms sink into the water. The platforms are formed by pouring concrete into platform-shaped forms, *lift* by *lift*—that is, one 20-ft level (or so) at a time. As the concrete cures, the structure is ballasted and lowered into the waters of a deep fjord. That way, work continues at the same elevation throughout the construction process, with only a few dozen feet of concrete showing above the water line.

The outer walls of a concrete platform might be 18 in. thick at the top, but as much as 3 ft thick at the bottom. The Troll concrete structure is located in 994 ft of water. It was installed in 1995 and remains today the deepest water gravity structure in the world. It weighs about 700,000 tons, with 110,000 tons being steel reinforcing rods. This "concrete" platform has 15 times the amount of steel as the Eiffel tower, and 1½ times the steel weight of the world's deepest fixed steel platform, *Bullwinkle*.

From Here to There

Moving a steel jacket from the construction yard to the drillsite requires a *launch barge*, a vessel specially suited—and sometimes specially built—for the job. During the *load out* onto the barge, the jacket is moved along skid ways onto the launch barge. Motive power comes from a series of hydraulic jacks pushing or winches with block and tackle pulling. The lateral push or pull force runs 5–10% of the total structure's weight. A 3000-ton jacket requires 150–300 tons of jacking or winching force. The agonizingly slow load out process typically takes up to a day—the bigger the jacket, the longer the day.

As the jacket moves onto the barge, the crew pumps water into some and out of other barge compartments, changing the buoyancy to accommodate the weight and position of the jacket. Once the jacket is properly positioned on the barge, it is elaborately welded to the barge using

Fig. 6–5 Concrete Gravity Structure being Towed to an Installation Site (Courtesy Norske Shell)

additional structural members. Losing a structure overboard could abruptly end an otherwise illustrious engineering career. Some jackets have been hauled as far as from a fabrication yard in Japan to the California coast.

Several seagoing towboats haul the barge/jacket combination to the vicinity of the drillsite. For *Bullwinkle* (*see* the photo in chapter 1, Fig. 1–16), a jacket that weighed 40,000 tons and required a launch barge 850 ft long and 180 ft wide, three tugs pulled and two tugs followed to provide braking, if needed. The capricious nature of inland waterways, and even some offshore locations, demands a careful depth survey of the route between the fabrication yard and the drillsite. A loaded launch barge can draw as much as 45 ft of water. Going aground also does little for an otherwise unblemished résumé.

A concrete structure already floats in the waters of a fjord and just needs a few towboats to haul it to the drillsite. The several hundred feet of platform that sit below the water's surface demand more towing power and breaking power than a launch barge. Typically, a concrete platform has to be raised some by partially de-ballasting to reduce the drag and to clear through shallower waters. Considering the massive size of these concrete monsters, it takes husky tugboats to tow them to location. Ten of the world's largest and most advanced tugs moved the Troll platform 174 miles (322 km) from its construction site to the final resting site. (*See* Fig. 6–5.) The vessels each had bollard pulls (power exerted on the platform's bollards, which are posts used to tie on hawsers) of 150–200 tons and ratings of 12,000–16,000 horsepower.

Installing Platforms

Conventional platforms and CTs

As the launch barge and its precious cargo near the drop site, the crew uses burning torches to remove the tie-down welds in preparation for launching the jacket. Hydraulic jacks or winches on the barge move the jacket slowly over the stern. The crew pumps the barge ballast around carefully, and at some point the jacket begins to slide off the barge on its own, as in Figure 6–6. Because of their hollow cylindrical shape, the members that make up the platform structure have about 6% net buoyancy, so the jacket floats on its side like a sunning whale. The launch crew opens valves on some of the cylindrical members, allowing water to enter. If they did their calculations right, the entire structure slowly rotates to an upright position with the top of the jacket poking 10–50 ft out of the water, enough for a few feet of clearance from the sea floor. The tugs tow the jacket over the drop site (guided by satellite positioning), the remaining floodable members are filled, and it slowly drops into place.

Fig. 6–6 Steel Jacket Launching from a Barge (Courtesy of Shell Exploration and Production Company)

Sometimes *mud mats*, large wooden or metal rugs, are placed under the bottom of the jacket to ensure it stays in place until the pilings can be driven to make a permanent connection to the sea floor. A transport barge with the steel piling delivers its cargo in short order and a crane lifts the pile sections, sometimes more than 200 ft in length, and slips them down *foundation sleeves* or slots in the jacket legs. Additional pile sections are welded on and driven into the seabed floor.

For onshore construction of large buildings, bridges, overpasses, etc., a common technique to achieve proper support is by driving steel, concrete, or even wooden piles into the soil. The resistance developed along the shaft of the pile is sufficient to carry the vertical, lateral, and overturning loads created during the various environmental and operational events above. The equipment and hardware used in the deepwater for installing platform foundations is more complicated, but the concept is similar. For the fixed platforms in less than 400 ft of water, the piles may be driven with surface pile drivers and long extensions to the pile being driven. Diesel, steam, or hydraulic fluid provide power to surface pile drivers whose mechanicals are exposed to the air around them. In the deepwater, pile extensions are impractical, so pile drivers sit totally underwater and completely enclosed. Hydraulic hammers drive the piles, controlled from a work boat at the water surface.

After a few piles secure each leg, the construction crew straightens the jacket by ballasting and de-ballasting some jacket segments and perhaps by having a crane hoist one side or another. The

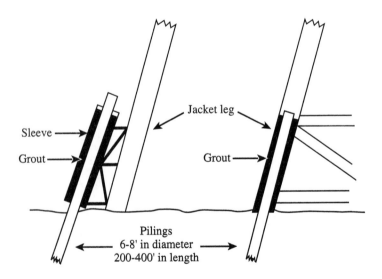

Fig. 6–7 Steel Jacket Connected to a Piling with Grout

crew then permanently attaches the jacket to the piles. They can either weld the piles directly to the jacket or attach them using cement grout. (*See* Fig. 6–7.) With the jacket now permanently secured, the rest of the piles are slipped into their sleeves, driven into place, and permanently attached to the jacket. That completed jacket awaits its deck to make it a complete platform.

Installing Concrete Gravity Platforms

A concrete platform arrives at the drop site already in the upright position. By slowly increasing the volume of water in the base by controlled flooding, the platform drops and pushes itself into the sea floor, perhaps several tens of feet, depending on the soil characteristics. The underside of the concrete base often has large structural ribs several feet deep. Seen from the inside, there is a rib extending around the perimeter, then some laterals creating cavern-like rooms. As the base contacts the sea floor, the foundation is pushed into the hard clay, much like a cookie cutter into dough. To avoid blowing out the water around the rib edges, valves on the top of the foundation can be opened to let the entrapped water out. Sometimes the water is pumped out to create downward suction force to aid in the leveling and installation.

As the huge structure is slowly and carefully lowered onto and into the sea floor, settling into the hard bottom is controlled to ensure that the base remains level and the columns vertical. Water can be added inside the structure to increase the weight; additional water can be pumped from the rib caverns to provide more downward suction forces. Once the structure has reached its required position and is within the specified tolerances for levelness, the base may be weighted with concrete or iron ore to achieve a predetermined on-bottom force. Because of the variations in bottom conditions, the detailed design for each gravity-based structure is different. Indeed in some areas of the world, the base penetrates very little into the sea floor. The subsoil conditions are extensively explored, and the appropriate design chosen for those specific conditions. For example, the Malampaya Gravity structure in the Philippines was set down onto a bed of pre-laid rocks, and then was weighted with iron ore. More rock was placed around the edges to control any tendency for scouring from currents.

Setting the Deck

If all goes well, the steel topsides deck for the steel jacket or compliant tower arrives atop a transport barge not long after the final integrity checks of the installed jacket. The same cranes

Fig. 6–8 *Saipem 7000* Heavy Lift Barge Setting a Deck Section (Courtesy Saipem)

used to set the jacket lift the deck into place, either as a complete unit or in sections. (*See* Fig. 6–8.) Pre-designed deck legs mate with the top of the jacket. Welding them together makes the final connection.

In many cases, the decks come complete with drilling and production equipment. That allows the crews at the fabrication yard to make all the pipe, electrical, hydraulic, or other connections. Alternatively, the equipment arrives piecemeal, and an army of welders, fitters, laggers, instrument technicians, and other craftsmen go aboard to install, connect, and prove readiness. Pre-installation is usually much cheaper, but the availability of cranes with enough capacity to lift the extra load can limit the choices.

A concrete structure is usually transported to the site upright, so the deck installation can take place in the fjord, before transport to the drillsite. *Floating the deck* over the concrete platform is the most common way to mate the two. In the fjord, protected from wind and waves, the concrete platform is ballasted so the top is almost at water level. One or two barges with the deck sitting on top float to a point where either the barges straddle the partially submerged concrete structure or the single barge moves inside the structure legs. By slowly ballasting the barges, the deck comes to rest on the platform, where the connections are made. Once the load transfers to the platform, the barges float away. Figure 6–9 shows the topsides for the Malampaya structure, offshore Philippines, being moved via barge into place to be transferred to the concrete structure sitting in the background. This particular mating took place at the final platform site after the hull was set on the bottom.

Fig. 6–9 Topsides Deck Ready to Be Lowered onto Legs of a Concrete Structure (Philippines Exploration BV)

Setting the Pipeline Riser

Sometimes the plan calls for a platform to tie into one or more export pipelines (inbound or outbound). As a last task, the crane vessel picks up the end of each pipeline for attachment to a riser, a vertical line attached to the platform and connected to piping on the deck. More on this follows in chapter 10 on laying pipelines. At this point, the operating crews can begin to drill wells or complete the predrilled wells and start production.

7

Floating Production Systems

On such a full sea are we now afloat,

And we must take the currents when it serves,

Or we lose our venture.

From *Julius Caesar*

William Shakespeare (1564–1616)

By their nature and training, explorers look for hydrocarbons wherever, and often that can lead them to the most inhospitable places. Practical development and production schemes may not always be high on their list of concerns. But drillers and production engineers continually rise to the occasion when explorers present opportunities that are continually deeper and otherwise less pleasant. Most often, they even encourage explorers to push on to the forbidding.

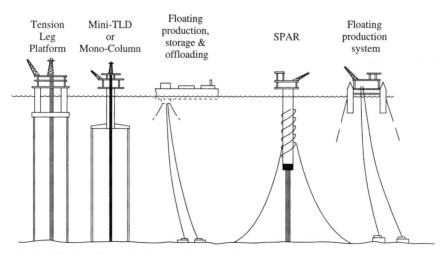

Fig. 7–1 Floating System Options for Deepwater Projects

Once offshore operations extended beyond practical fixed platform limits, the production engineers borrowed concepts devised by the drilling engineers. They in turn had responded to the needs of the explorers with semisubmersibles and drillships as they moved out of shallow water. Thus floating production systems (plus, in many cases the subsea completions covered in the next chapter) now provide the viable options in deepwater.

FPS's come in many sizes and shapes. (*See* Fig. 7–1.) Some provide more functions than others. In every case, they differ from fixed systems by what holds them up—the buoyancy of displaced water, not steel understructure.

Floating systems have four common elements:

Hull – the steel enclosure that provides water displacement. Floating system hulls come in ship shapes, pontoons and caissons, or a large tubular structure called a *spar*.

Topsides – the deck or decks have all the production equipment used to treat the incoming well streams plus pumps and compressors needed to transfer the oil and gas to their next destinations. Some have drilling and workover equipment for maintaining wells. Since almost all deepwater sites are somewhat remote, their topsides include living accommodations for the crew. In most cases, export lines connect at the deck also. More discussion of the deck and its equipment follow in chapter 9.

Mooring – the connection to the seabed that keeps the floating system in place. Some combine steel wire or synthetic rope with chain, some use steel tendons. In some cases, they make a huge footprint on the seabed floor. (*See* Fig. 7–2.)

Fig. 7–2 A Rendition of the Tension Leg Platform, *Auger*, Sometimes Called "The TLP that ate New Orleans" (Courtesy Shell Exploration and Production Company)

Risers – steel tubes that rise from the sea floor to the hull. A *riser* transports the well production from the sea floor *up* to the deck. The line that moves oil or gas in the other direction, from the deck *down* to pipeline on the sea floor, uses the oxymoron export risers. Chapter 10 covers risers in more detail.

Tension Leg Platforms (TLP)

The semisubmersible, used for years only for drilling, begat TLPs. By similar design, the buoyancy of a TLP comes from a combination of pontoons and columns. (*See* Fig. 7–3.) Vertical *tendons* from each corner of the platform to the sea floor foundation piling hold the TLP down in the water. Vertical risers connected to the subsea wellheads directly below the TLP bring oil and gas to dry trees on the deck.

The vertical tendons are steel tubes, typically with 24- to 32-in. diameters with wall thicknesses of 1–2 in. With these tendons under constant tension, there is very little vertical movement at the deck, even in heavy weather. However during a *design storm*, the worst wind and sea conditions the designers can imagine, the TLP deck moves laterally as much as 8% of the water depth—within a radius of 240 ft in 3000 ft of water. Drillpipe can accommodate the lateral movement that comes from moderate winds or ocean current changes. In heavy seas that push the TLP towards its limit, no drilling takes place anyway.

Installation. While a TLP is being constructed at an onshore construction yard, a floating drilling rig is often busily drilling several of its wells at the offshore location. The drilling operation leaves behind subsea wellhead assemblies which are each later connected to the production risers.

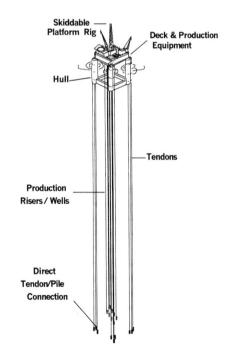

Skiddable Platform Rig

Deck & Production Equipment

Hull

Tendons

Production Risers / Wells

Direct Tendon/Pile Connection

Fig. 7–3 TLP for the *Brutus* Field in the Gulf of Mexico (Courtesy Shell Exploration and Production Company)

Sea-going tugs tow the TLP hull to the site. Crane barges are used to assemble the steel tendons that connect the TLP to preinstalled anchor systems on the bottom. With the TLP secure, the drilling rig onboard has to connect (or re-enter) the pre-drilled wells by attaching a riser (or connector pipe) to the subsea wellheads, one at a time. Since the TLP is not directly above each well, a small structural frame on wire lines is lowered and connected to the subsea wellhead. A *remote operated vehicle* (ROV) guides the frame into place and manipulates the locking device to secure it. The riser (or connector pipe) then uses the wire lines as a guide to the wellhead, where an ROV assists in the mechanical connection.

Without being directly above the wellhead, how does the riser achieve a square, tight fit? As a frame of reference, imagine a 10-ft steel rod, say with 1/8-in. diameter, hanging from the ceiling. Moving the bottom of the rod sideways 10 in. is the same scale as the 240-ft offset for the tendons and risers in 3000 ft during a design storm. Bending the 10-ft rod to make a connection at the bottom deforms its shape imperceptibly. Not easy, but doable. Even better, pushing the bottom of the rod only a couple of inches is analogous to the maneuvering the ROV deals with during the alignment process.

Fig. 7–4 TotalFinalElf *Matterhorn SeaStar©*, a Monocolumn TLP (Courtesy Atlantia Offshore, Ltd.)

Dry trees on the deck of the TLP control the flow of oil and gas production coming up through the conductor pipes. However, like other floating systems, it can receive production from risers connected to remote subsea wet tree completions. Most TLPs have *subsea riser baskets*, structural frames than can hold the top end of risers coming from subsea completions.

Monocolumn TLP

In shallower water or for smaller deposits in deepwater, and where no more drilling is planned, some companies use a smaller variation of the TLP called a mini-TLP, a *monocolumn* TLP, or sometimes a *SeaStar*. The names *monocolumn* and *SeaStar* (a proprietary label) come from the underwater configuration of the flotation tanks, a large central cylinder with three star-like arms extending from the bottom. (*See* Fig. 7–4.) The cylinder measures about 60 ft in diameter and 130 ft in height. The arms reach out another 18 ft.

As with other TLPs, tendons secure the substructure to the sea floor, in this case two from each arm. The mooring system, risers, and topsides are similar to any other TLP, except for the modest sizes. The absence of drilling equipment on board helps lower the weight of the topsides and allows this scaled-down version.

FPSO

From 400 yards away, most FPSOs are indistinguishable from oil tankers. In fact, while many FPSOs are built from scratch, the rest are oil tankers converted to receive, process, and store production from subsea wells. FPSOs do not provide a platform for drilling wells or maintaining them. They do not store natural gas, but if gas comes along with the oil, facilities onboard an FPSO separate it. If there are substantial volumes, they are sent back down a riser for reinjection in the producing reservoir or some other nearby home. (*See* Fig. 7–5.)

Fig. 7–5 The FPSO *Anasuria* in the North Sea (Courtesy Shell International, Ltd.)

Shell installed the first FPSO, a tanker conversion, in 1977 to produce from the small Castellon field in the Mediterranean Sea. Since then, the industry has found scores of remote or hostile environments that call for the FPSO design:

- at sea where no pipeline infrastructure exists
- where weather is no friend, such as offshore Newfoundland or the northern part of the North Sea
- close to shore locations that have inadequate infrastructure, market conditions, or local conditions that may occasionally not encourage intimate personal contact, such as some parts of West Africa.

As an FPSO sits on station, wind and sea changes can make the hull want to *weathervane*, turn into to the wind like ducks on a pond on a breezy day. As it does, the risers connected to the wellheads, plus the electrical and hydraulic conduits, could twist into a Gordian knot. Two approaches deal with this problem, the cheaper way and the better way.

Fig. 7–6 An FPSO Internal Turret (Courtesy FMC SOFEC Floating Systems)

Fig. 7–7 An FPSO Cantilevered Turret (Courtesy FMC SOFEC Floating Systems)

In areas of consistent mild weather, the FPSO moors, fore and aft, into the predominant wind. On occasions, the vessel experiences quartering or broadside waves, sometimes causing the crew to shut down operations.

In harsher environments, the more expensive FPSOs have a mooring system that can accommodate weathervaning. Mooring lines attach to a *revolving turret* fitted to the hull of the FPSO. As the wind shifts and the wave action follows, the FPSO turns into them.

The turret might be built into the hull or cantilevered off the bow or stern. (*See* Figs. 7–6 and 7–7.) Either way, the turret remains at a permanent compass setting as the FPSO rotates about it.

The turret also serves as the connecting point between the subsea systems and the topsides production equipment. Everything between the seabed and the FPSO is attached to the turret— production risers, export risers, gas reinjection risers, hydraulic, pneumatic, chemical, and electrical lines to the subsea wells, as well as the mooring lines.

Turrets contain a *swivel stack*, a series of fluid flow and electronic continuity paths that connect the seaside lines with the topsides. As the FPSO swings around the turret, the swivels redirect fluid flows to new paths, inbound or outbound. Other swivels in the stack handle pneumatics, hydraulics, and electrical signals to and from the subsea systems.

In some designs, the FPSO can disengage from the seabed (after shutting in the production at the wellheads) to deal with inordinately rough seas, or other circumstances that might worry the ship's captain, like an approaching iceberg. A *spider buoy*, the disconnectable segment of the turret with the mooring, the riser, and the other connections to the subsea apparatus, drops and submerges to a pre-designated depth as the vessel exits the scene.

The full turret aboard the PetroCanada's FPSO *Terra Nova* weighs more than 4400 tons and is 230 ft high. The turret assembly includes a spider buoy, 65 ft in diameter and 1400 tons, with connections for 9 mooring lines and 19 risers. On disconnect, which takes only 15 minutes, the mooring lines, risers, and conduits remain tethered to the spider buoy, until the FPSO returns and fishes it out of the water, reconnects, purges the lines, and restarts operations.

After the oil moves from the reservoir to the FPSO via the turret, it goes through the processing equipment and then to the storage compartments. The 14 storage tanks aboard the *Terra Nova* range from 50,000 to 78,000 barrels each, with total vessel capacity of 960,000 barrels. The largest measures 88 by 56 by 85 ft high. Like all new oil tanker construction, the *Terra Nova* has a double hull. The oil storage sits in the inner hull, separated by air space from the outer hull. That reduces the risk of an oil spill should an accident result in piercing the outer hull.

Shuttle tankers periodically must relieve the FPSO of its growing cargo. Some FPSOs can store up to two million barrels on board, but that still calls for a shuttle tanker visit every week or so. Mating an FPSO to the shuttle tanker to transfer crude oil calls for one of several positions.

- The shuttle tanker can connect to the aft of the FPSO via a *mooring hawser and offloading hose.* The hawser, a few hundred feet of ordinary marine rope, ties the shuttle tanker to the stern of the FPSO, and the two vessels weathervane together about the turret.
- The shuttle tanker can moor at a buoy a few hundred yards off the FPSO. Flexible lines connect the FPSO through its turret to the shuttle buoy and then to the shuttle tanker.
- Some shuttle tankers have dynamic positioning, allowing them to sidle up to the FPSO and use their thrusters on the fore, aft, and sides to stay safely on station, eliminating the need for an elaborate buoy system. The shuttle tanker drags flexible loading lines from the FPSO for the transfer.

Oil then flows down a 20-in.(or so) offloading hose at about 50,000 barrels per hour, giving a turnaround schedule for the shuttle tanker of about one day.

FDPSO

Inevitably, the attributes of a FPSO plus a drillship or a semisubmersible come together in the form of a floating drilling production storage and offloading (FDPSO) to do it all in the deepwater. The crucial technologies for this union, still in the nascent stage, are motion and weathervane compensation systems for the drilling rig.

FSO

This specialty vessel stores crude from a production platform, fixed or floating, where no viable alternatives for pumping oil via pipeline exist. FSOs almost always had a former life as an oil tanker and generally have little or no treating facilities onboard. As with the FPSO, shuttle tankers visit periodically to haul the produced oil to market.

FPS

In theory, an FPS can have a ship shape or look like a TLP, with pontoons and columns providing buoyancy. Almost all are the TLP type design. Either way, the FPS stays moored on station to receive and process oil and gas from subsea wet trees, often from several fields. After processing, the oil and gas can move ashore via export risers, or the gas can go into reinjection and the oil to an FSO.

At a typical operation, the Shell-BP *Na Kika* project in the Gulf of Mexico, the FPS is designed to handle six oil and gas fields, some several miles away. (*See* Fig. 7–8.) Production arrives at the FPS from subsea completions through flexible and catenary risers (more on this in chapter 10 on pipelines and risers) and goes through treating and processing before it leaves via export risers towards shore. The FPS provides a home for the subsea well controls that are connected via electrical and hydraulic umbilicals.

Petrobras led the industry up the FPS learning curve in the 1990s. They developed most of their deepwater fields with subsea wellheads, using FPS's as nodes on a gathering system, and subsequently pumping the oil and gas onshore or the oil to a nearby FSO.

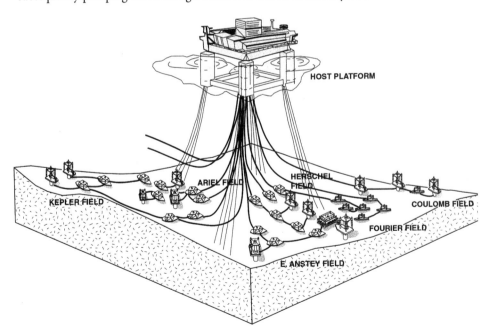

Fig. 7–8 The *Na Kika* Development Scheme Uses an FPS (Courtesy Shell Exploration and Production Company)

Spars

Even though the name *spar* comes from the nautical term for booms, masts, and other poles on a sailboat, spars exhibit the most graceless profile of the floating systems. An elongated cylindrical structure, up to 700 ft in length and 80 to 150 ft in diameter, the spar floats like an iceberg—it has just enough freeboard to allow a dry deck on top. (*See* Fig. 7–9.) The mooring system uses steel wire or polyester rope connected to chain on the bottom. The polyester has neutral buoyancy in water and adds no weight to the spar, eliminating having to build an even bigger cylinder. Because of its large underwater profile, the huge mass provides a stable platform with very little vertical motion. To ensure that the center of gravity remains well below the center of buoyancy (the principle that keeps the spar from flipping), the bottom of the spar usually has ballast of some heavier-than-water material like magnetite iron ore.

Because of the large underwater profile, spars are vulnerable not only to currents but also to the vortex eddies that can cause vibrations. The characteristic *strakes* (fins that spiral down the cylinder), apparent in Figure 7–10, shed eddies from these ocean currents, although the strakes add even more profile that calls for additional mooring capacity.

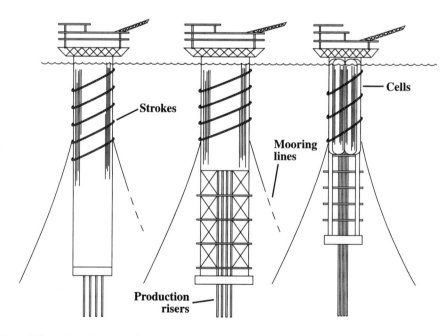

Fig. 7–9 Three Spar Options—Conventional, Truss, and Cell

Fig. 7–10 The *Genesis* Spar en Route to the Wellsite (Note the spiral strakes; courtesy ChevronTexaco)

Drilling rigs operate from the deck through the center of the cylinder. Wells connect to dry trees on the platform by risers, also coming through this core. Risers from subsea systems and export risers also pass through the center.

Spars have evolved through several generations of design. (*See* Fig. 7–9.) The original concept had a single 600-ft steel cylinder below the surface.

Not long after that, the *truss spar* arrived with three sections: a shortened "tin can" section; below that, a truss frame (saving weight); and below that, a keel or ballast section filled with magnetite. The truss section has several large, horizontal, flat plates that provide dampening of vertical movement due to wave action. Like the original, the cylindrical tank provides the buoyancy for the structure and contains variable ballast compartments and sometimes tanks for methanol, antifreeze used to keep gas lines from plugging.

The third generation, the *cell spar*, is a scaled down version of the large truss spar and is suitable for smaller, economically challenged fields. The design takes advantage of the economies of mass production. It uses more easily fabricated pressure vessels, what refineries and gas plants call bullets that are used to handle volatile hydrocarbons. Each vessel is 60 to 70 ft in diameter and 400 to 500 ft long. A *cell*, a bundle of tubes that looks like six giant hot dogs clustered around a seventh, makes up the flotation section extending below the decks. Structural steel holds the package together, extends down to the ballast section, and can include heave plates.

Spars have also served solely as oil storage vessels, which of course made them FSOs. The most notorious, the Brent Spar, had a useful life until the Brent platform in the North Sea gained full access to an oil export system. The Brent Spar disposal saga beginning in 1995 became an epoch lesson in business/societal relations. The spar owner's plan to sink it in the deep Atlantic was aborted when Greenpeace landed a boarding party of squatters. Ensuing international protests, which included firebombing gas stations in Germany, led to postponed disposal, lengthy negotiations, and shoreside dismantling years later.

Construction and Installation

Ship-shape hulls. FPSOs, FSOs, and the ship-shape FPS's present no inordinate technological challenges to shipyards around the world, with the exception of the turret systems. Consequently, turrets are usually built by specialty constructors and moved to the shipyards for installation.

Like all ship-shape hulls nowadays, construction takes place in a dry dock on blocks with the usual list of craftsmen—steelworkers, welders, pipe fitters, laggers, and electricians. When the hull reaches satisfactory integrity, the dry dock is flooded and the hull is floated to a dockside. Equipment installation and outfitting continues, especially the turret or special anchor points and winches for a mooring system.

Caisson and pontoon hulls. TLP and FPS hulls look much alike, but the mooring details differ. Both have large volume pontoons and columns (or caissons) that provide buoyancy to carry the deck, riser, and mooring line loads.

Most FPS hulls are built in construction yards rather than shipyards, avoiding the need for expensive dry docks. The physical shape of the finished product allows the steel work and assembly to take place on a flat spot in the yard. These behemoths use thousands of tons of steel and require several hundred craftsmen for months. The Shell *Ursa* TLP, built in Italy and installed in the Gulf of Mexico in 3800 ft of water, has four 85-ft diameter columns, each 177 ft tall, connected with four ring pontoons, each 38 ft wide and 29 ft deep. The hull alone weighs 28,000 tons. Total steel, including the deck, hull tendons, and foundation piling adds up to 63,000 tons. Large mobile cranes handle the pieces as the overall structure comes together. Figure 7–11 shows a TLP hull under construction in Italy.

Fig. 7–11 TLP Hull under Construction (Courtesy Shell Exploration and Production Company)

Spars. Construction of a spar requires similar yard facilities to the TLP and FPS hulls. The design of the flotation cylinder requires internal stiffening pieces to accommodate the hydrostatic forces from the surrounding water and to carry the deck and mooring loads.

Spars are also massive structures. The ChevronTexaco Genesis hull is 122 ft in diameter and 705 ft deep and used 26,600 tons of steel. It was fabricated in two sections in Finland then carried to the Gulf of Mexico on a large lift vessel. The hull incorporates a center well bay that is 58 by 58 ft and accommodates 20 wells.

From here to there

The floating systems are built, of course, many miles from their final destination. For example, most of the TLP hulls for the Gulf of Mexico came from Italy and Korea. Finland specializes in spar hulls. Ship-shape hulls come from all over the world.

For the ship-shape hulls, transportation is straightforward—literally. The hull either drives itself to the location (if it has its own engine power), or sea going tugs tow it. Since their mission is more or less stationary, most FPSOs and FPS's do not have their own motoring facilities.

The pontoon/caisson hulls move to the wellsite either by *wet tow* or *dry tow*. For a wet tow, tugboats drag the hull at an agonizingly three to four knots. Transport from Italy to the Gulf of Mexico takes about 90 days.

The dry tow method involves placing the hull on a vessel with heavy lift capability and a more nautical-friendly shape, sometimes self-powered, sometimes towed. In either case, speeds of 10 to 12 knots cut the Italy/Gulf trip to about 25 days.

The quayside water is not likely to be deep enough to float a pontoon/caisson hull. It may be deep enough so that the hull can be dragged out of the construction yard on skids right on to a transport vessel. If the waterway is not deep enough, shallow draft transit barges might have to be pressed into service to carry the hull to deeper, but still protected, waters. At that point, the barges are lowered by ballasting until the hull floats. The larger, deeper draft, heavy lift vessel is ballasted enough to come under the hull, where it is de-ballasted, lifting the hull. Figure 7–12 shows a TLP hull en route on a heavy lift vessel.

Fig. 7–12 Hull of TLP *Brutus* en Route to Installation on a Heavy Lift Vessel (Courtesy of Dockwise USA)

Spars move more or less the same way, though some are transported in two or three pieces, with final assembly near the drop site.

Installation

If all goes well, by the time a floating system arrives on site, the anchoring system is in place. In the case of an FPS, FPSO, FSO, and spar, mooring lines and chain have also been installed and are waiting on the sea floor. A large crane vessel fishes the mooring lines out, transfers them to the hull, one at a time. The lines are winched to the right tension and position and locked in place.

For a TLP, the tendons arrive at the site in 200-ft segments. The first segment is hung off the side of the construction vessel, and the next one is mechanically connected to it. Those two are then lowered into the water and hung off, and the process is repeated until the bottom segment reaches its destination, a specially designed receptacle on a pile on the sea floor. For the *Ursa* TLP, each 3800-ft segment has a 32-in. diameter with 1.3-in. thick walls and weighs 1000 tons. Each tendon connects to a separate pile that has a diameter of 96 in. and has been driven 400 ft into the seabed.

To complete the installation, the TLP, ballasted below normal depth, is moved over the site, and the tops of the tendons are secured to the TLP keels. Then the TLP is de-ballasted, causing

Fig. 7–13 The *Genesis* Spar Uprighting from Its Floating Position (Courtesy ChevronTexaco)

tension in the tendons as it rises. The tension can reach 1000–2000 tons per tendon ensuring that the tendons never go slack during any environmental or operational changes. Otherwise, they could bend or buckle, leading in the extreme to the frightening specter of a TLP bobbing up and down like a cork.

Hauling a 600- or more ft spar out to sea sometimes calls for the fabrication yard to build it in two pieces, tow them to the site, and weld them together. Once connected, the crews carefully flood the spar's compartments so that it rotates to the upright position, mostly submerged. (*See* Fig. 7–13.) Heavy lift cranes resurrect the mooring system and assist with the connections.

Setting the deck

Expertise and craftsmen for hulls and decks are different enough that the decks for TLPs, spars, or FPS's usually are constructed at separate yards than the hull. The hull and the deck can be mated onshore at an outfitting and commissioning site or at sea. Either way, the deck can come whole or in several pieces.

At sea, heavy lift cranes can place the deck or its pieces on top of the moored hull. Alternatively, the *float-over* method can be used. The deck on two parallel barges is floated over a ballasted floater hull. The water is pumped out of the floater, allowing it to rise and pick up the deck from the float-over barges.

Most FPS's, FPSOs, and TLPs have their topsides deck equipment set at an onshore or near onshore location. Spars, having their awkward 600 ft height, always have their decks set after turning upright on site. Heavy marine lift vessels do the transfer of the topsides, which usually come in one unit.

Mooring Spreads

A typical mooring configuration for FPSOs, FSOs, and spars has 4-in. rope, steel wire or polyester, stretching down to heavy chain links, each weighing more than 500 pounds (Fig. 7–14). The chains connect at the seabed to an anchor system, usually made up of steel piling. These *foundation piles* can be driven into the seabed by underwater hydraulic hammers. Alternatively, they can be installed by the *suction pile* method.

Fig. 7–14 A 500-Pound Chain Link for a Mooring Chain (Courtesy Offspring International, agent for Zheng Mao Group Ltd.)

Like rubber boots stuck in the mud, suction piles rely on the sediment around them to hold them in place. The installation is a bit more complicated than donning footwear. Suction pile sizes vary, but the *Na Kika* project piles and installation serve as an example. A 14-ft diameter tubular about 80 ft long, with a top cover was lowered to the sea floor. As the bottom edge touched the soil, the enclosed water escaped through valves left open on the top. The weight of the steel suction pile was enough to cause the pile to penetrate about 40 ft into the soil. As it penetrated, soil moved up the tubular, pushing the water out the top, avoiding any water escaping around the bottom. Otherwise the soil in the area might have washed out, causing a loss in load-carrying capacity.

After about 40 ft, the tubular weight was not enough to cause further penetration. A pump, operated via an ROV, sucked more water from the top of the tubular, causing downward suction that caused the piling to move slowly down for the full 80 ft. The valves were closed, and the piling was permanently in place, held by suction akin to a rubber boot stuck in 80 ft of mud.

Mooring spreads have at least 8 separate legs coming from the floating system; some have up to 16. Like any stable anchoring system at sea, the length of the anchor line runs one and a half to two times the water depth. For the FPS used in the *Na Kika* project in 6500 ft of Gulf of Mexico water, Shell/BP used 9600 ft of wire and 1750 ft of chain for each of 16 separate legs. The suction piles sit a mile and a half from the FPS. (*See* Fig. 7–15.)

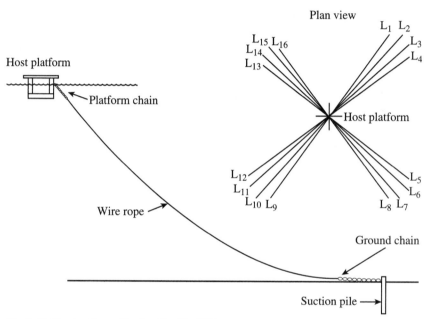

Fig. 7–15 Mooring Spread for the *Na Kika* FPS

Service companies install the foundation piles and the rope and chain before the arrival of a floating system, leaving the ropes on the seabed (in satellite-identified locations so they can find them) until hook-up time arrives. Then they use global positioning systems again to find the loose ends. With the assist of ROVs, they attach the ends to heavy-lift cranes that hoist them to the connect points on the floater.

Risers

FPS's connect to subsea wells by a combination of production risers and flowlines along the bottom. Chapter 10 treats them in detail.

8

Subsea Systems

Below the thunders of the upper deep

Far, far below in the abysmal sea

From *The Kraken*

Alfred, Lord Tennyson (1809–1892)

People sit in climate-controlled offices surrounded by multicolored electronic displays, control panels, and communication links. A transient signal crosses a screen and an un-ignorable alarm seizes the attention of an operator who punches changes into keyboard *in staccato* and then watches the screen intently.

Is this the Johnson Space Center? The Pentagon Command Center? No, it's a control room aboard a deepwater platform where oil and gas from subsea wells, a few miles out and a few thousand feet below on the sea floor, is managed. Advancing subsea technology now approaches the sophistication of outer space and military surveillance.

Companies already use subsea systems to tap oil and gas fields in two ways. First, they connect smaller fields to existing infrastructure, obviating the killer cost of a brand new platform. Figure 8–1 shows the arrangement for the *Angus* field in the Gulf of Mexico, where the existing *Bullwinkle* infrastructure is effectively used. Second, subsea systems also have a place where no infrastructure exists. A combination of smaller fields, close to each other but not reachable by directional drilling and each not large enough to support its own platform, can be developed with a subsea system and a new, common platform. The *Na Kika* system, also in the Gulf of Mexico, mentioned in chapter 7 and shown again in Figure 8–2, is such a multi-field development.

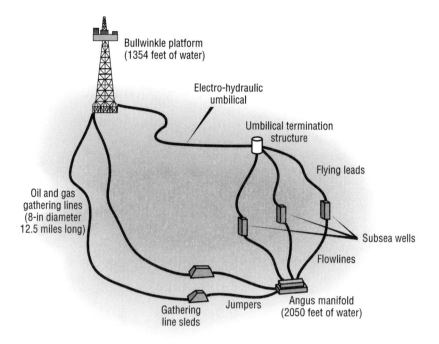

Fig. 8–1 Subsea Development Scheme for *Angus*, Hooked up to *Bullwinkle* (after INTEC)

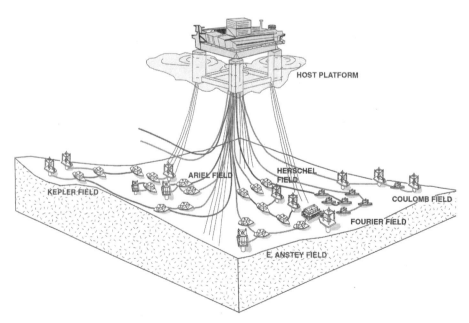

Fig. 8–2 The *Na Kika* Development Scheme for Six Subsea Fields (Courtesy Shell Exploration and Production Company)

Six main components make up a subsea system.

- Wells
- Subsea trees
- Manifolds and sleds
- Flowlines
- Electric and hydraulic umbilicals
- Subsea and surface controls

In addition, these components are connected by jumpers and flying leads.

Wells

The designs and specifications of all the subsea components—trees, manifolds, umbilicals, and so on—are a function of the characteristics of the well. For example, some subsea wells have high shut-in pressures, greater than 10,000 pounds per sq in., while others are less than 5000 lbs per sq in. Some, with very low reservoir pressure, may require downhole pumps to move the fluids to the surface. The functions of the trees and other downstream components are similar in all cases, but the hardware and various control devices differ in strength, materials, controls, instrumentation, and installation. Nonetheless, the pieces fit together as a system to accommodate the reservoir characteristics and the well parameters.

Trees

Subsea trees sit on top of the well at the sea floor. Although they have little visual resemblance to the original onshore Christmas trees, they provide essentially the same functions. They furnish the flow paths and primary containment for the oil and gas production and the valves needed for both operation and safety. Operators on the host platform actuate hydraulically controlled valves on the trees and monitor feedback from sensors on the tree and in the well.

Remote and inaccessible as subsea trees (and other subsea equipment) are, they demand robust design, built-in redundancy, vigilant manufacturing quality control, and thorough testing of the finished product to reduce some worry of equipment failure. Even then, nothing is perfect, and components may have to be disconnected and brought back to the surface for repair.

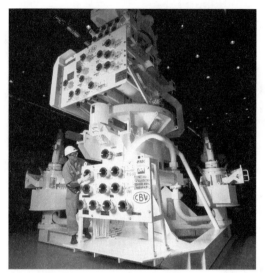

Fig. 8–3 A Subsea Tree with ROV-Friendly
Connections (Courtesy FMC Technologies)

Subsea trees normally have the external handles and fixtures evident in Figure 8–3. They enable ROVs to physically turn valves and activate other control functions, sometimes during normal operations, other times as a backup if something fails. An important ROV capability is to provide hydraulic pressure to open or close various valves in the event a primary electro-hydraulic system misfires. In addition, ROVs hook up, disconnect and otherwise change around control lines. They also install rigging and do other miscellaneous maintenance activities. And as important as anything, ROVs provide visual confirmation that what was supposed to happen happened. Design of subsea components takes all these factors into consideration.

Manifolds and Sleds

A *manifold* is quite simple in concept—it provides the node between the individual flowlines from the wells and the flowline to the host platform. A *sled* is a termination structure for a flowline or gathering line on the one side and a connection to a subsea well or manifold on the other. (*See* Fig. 8–1.)

In the simplest case (a single well development), the oil and gas production moves from the subsea tree via a short *jumper* to a flowline sled, into a flowline and then directly to the host platform. In most fields though, several wells are looking for a way to the surface. In those cases, several flowlines converge and are connected to a central subsea manifold with jumpers. Sometimes several wells are clustered around and linked directly to a common manifold. Using sleds, jumpers, and other termination structures allows major work modules to proceed on independent work schedules. Installing wells and manifolds and laying pipe can happen as vessels and equipment become available, a significant cost saving.

The manifold collects the produced fluids from individual wells, commingles them in a common *header* (sometimes referred to as a wide spot in the line) and sends them on to the host platform. For example, several 6-in. flowlines from individual wells connect to a length of header pipe with a 10- to 12-in. diameter. The commingled flow would then go through a gathering line to the host. In many cases, manifolds have extra slots for future wells or for tie-ins from other fields that might not warrant full subsea development on their own.

Manifold and sled designs vary in sophistication and complexity. Some are "dumb" steel structures that anchor the flowlines and collect the produced oil and gas for transport to the host platform. Others have complex control and distribution systems. Those give operators located at host platforms some miles away the ability to isolate flow from individual wells or fields and provide well flow data for reservoir surveillance and management.

Flowlines, Jumpers, and Gathering Lines

A whole chapter on pipelines and flowlines follows this one, but for completeness, a brief explanation is useful here. The oil and gas flows from the wells to a manifold via a *jumper* if it is close, say 50 ft or so, or through a flowline if not. A jumper is a prefabricated section of steel pipe specially configured to make a specific connection or it is a length of flexible composite line.

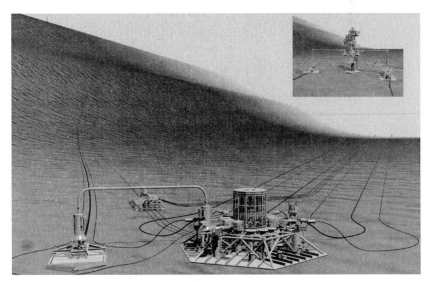

Fig. 8–4 The *Mensa* Subsea Field Manifold, Jumpers, Flying Leads, and Sled (Courtesy Shell Exploration and Production Company)

The precise measurements that determine the jumper shape—length, orientation, relative offset, and angles between the termination points can only be taken after the sleds, wells, and manifolds are in place.

The *Mensa* field, a high-pressure dry gas development in 5400 ft of Gulf of Mexico water (Fig. 8–4), uses short jumpers from each tree to individual well gathering lines. Three separate lines then proceed about five miles to a common manifold. From there, the commingled production is transported via a common flowline to the *Bullwinkle* platform 58 miles away for treatment and processing.

Umbilicals and Flying Leads

Like their namesakes in the animal kingdom, *umbilicals* conduct the flows necessary to keep the system alive. They provide the connecting medium for electrical, hydraulic, chemical injection, and fiber optic connections between the topsides facilities on the host platform and various subsea items—the manifold, sleds, termination structures, subsea trees, and controls. The number and the character of these umbilicals vary according to specific system needs and development plans.

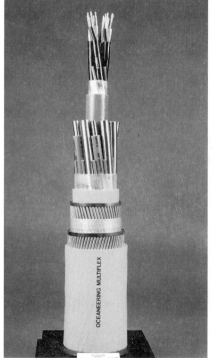

Umbilicals may be single function, e.g., a hydraulic line only, but they are more commonly multi-function *integrated umbilicals*, which provide hydraulic tubes, electrical lines, and tubes that carry chemicals to the manifolds and trees. Figure 8–5 shows a cutaway section of an 8-in. diameter full-service umbilical.

Some steel tubes in the outer ring can be used to pump chemicals to the production stream. Others can deliver hydraulic fluid to actuate subsea valves. Some umbilicals have thermoplastic tubing for low-pressure chemical injection service.

Fig. 8–5 Cutaway View of an Electro-Hydraulic Umbilical
(Courtesy Oceaneering Multiflex)

The center of the umbilical in Figure 8–5 has several electrical conductors that transmit signals from instrumentation on the subsea components (temperature, pressure, integrity checks) back to the control center. Electrical power can move through an umbilical to operate solenoid valves on the subsea control pods. In turn, those control hydraulic pressure supplied to the subsea valves on the manifolds, trees, or sleds. In some cases, power is also supplied to operate wellbore or subsea pumps.

Combining multiple services into a single umbilical can save manufacturing and installation costs. However, distance, depth, and weight may require multiple umbilicals. In the *Na Kika* development, several umbilicals lie on the sea floor between the host platform and the various trees, manifolds, and sleds.

The subsea end of an umbilical has a termination head that connects to a distribution structure, the *Umbilical Termination Structure* or *UTS*. (Sometimes the connection piggybacks instead on a manifold.) From there, the umbilical services are distributed to trees or any other miscellaneous equipment in the vicinity by *flying leads*, sort of subsea extension cords that are "flown" by ROVs and plugged into waiting receptacles. However, simple as it sounds, the umbilical cord and plug are considerably more sophisticated than a rubber-covered copper wire with a two-prong connection.

Control Systems

The ability to monitor and control wells and manifold functions from the host facility with confidence and safety is critical to overall subsea system performance. Trees and manifolds have *control pods*, modules that contain the electro-hydraulic controls, logic software, and communication signal devices. Redundancy is built into the systems to provide alternate ways to retrieve and transmit data and commands. Most control pods are designed so that if they fail, they can be replaced. Collaborating with a surface vessel that has an onboard winch, an ROV can fly in, disconnect the pod from its support structure, and connect the lifting slings so the winch can pull the pod to the surface. This is more or less a subsea version of changing a card in a computer.

The master computer in the host platform control room communicates with the subsea control pods. The pods then operate the valves and other functions on the manifold to increase or reduce flow rate, or to shut in the flow entirely, if needed. In addition, the system contains failsafe devices to shut in automatically if certain parameters are exceeded, such as large increases in well pressure that might indicate a plug in the system or an abrupt pressure drop that might indicate a flowline or other component failure. As a sometimes last but slow resort, an ROV can fly down and mechanically open and close valves or perform other functions.

For normal (and predominant) operations, the control room operators monitor and tweak the system based on feedback from the electronic monitoring devices, just as they would for the more accessible dry tree systems.

Flow Assurance

Subsea system designers and operators have *plugged lines* at the top of their worry list. Subsea facilities sit in the sunless deep in temperatures of 30–36 °F. Well fluids running through the flowlines, jumpers, and gathering lines to the host quickly lose heat to the surrounding environment. The cooling problem is magnified if the production system must be shut-in for any length of time greater than two to three hours. In both oil and gas wells, the water present in the produced fluids can form natural gas hydrates, a crystallized form of water and methane stabilized by high pressure. Hydrates can form at temperatures as high as 80 °F, depending on the pressure, in typical deepwater oil and gas producing systems.

At low temperatures, the waxy paraffins in some crude oils deposit on the pipeline wall, constricting flow, even blocking the flow entirely. To mitigate the temperature drop, the flowlines and other subsea components have insulation. The most effective insulation system, used in cases where there is a high potential for constriction or plugging, is the *pipe-in-pipe* flowline. The space between the pipes is partially evacuated and then filled with insulating material such as syntactic polypropylene. The result behaves like a super-elongated thermos bottle, eliminating most of the heat loss from the well fluids. These efficient insulation systems allow the well fluid to be transported great distances while maintaining almost all their original temperature, preventing the formation of both hydrates and wax.

From time to time, subsea facilities—like any other oil- and gas-producing facility—have to be shut-in. During that time, non-flowing, stagnant fluids sit in the flowlines and gathering lines. Even pipe-in-pipe insulation doesn't prevent flowline cooling indefinitely. Several methods deal with the long shut-in challenge.

- Slugs of chemicals can be injected into the production stream just prior to the shut-in, using either the umbilical or a separate chemical injection line. Methanol or glycol in amounts of

25–50% of the water content serve as antifreeze by depressing the freezing point below the seabed temperature, preventing hydrate formation.

- Where the water content of the gas is high, methanol or glycol injection is expensive. In that case, operators may depressurize the flowlines, moving the ambient conditions out of the hydrate-forming zone.
- The flowlines can be purged with dry oil pumped down through an umbilical-type connection.
- The flowline can be electrically heated. One electrical heating system now in use in the Gulf of Mexico uses the two pipes in a pipe-in-pipe flowline as the circuit that carries the electric current needed for heating.
- *Paraffin inhibitors* (chemical additives) can be injected into the well stream to keep paraffins and waxes from solidifying or depositing on the pipe wall.

Besides paraffins and hydrate plugging problems, deposition of inorganic scales, precipitation of organic molecules known as asphaltines, and corrosion problems can undermine flow assurance. The solution to these irritations is usually chemical additives. While effective, operators have found that these chemicals are expensive and take extensive adjustment to ensure continuous, problem-free, minimum-cost flow.

System Architecture and Installation

In the early days, subsea development systems were built with rigid, hard-plumbed templates. They were installed and connected in a given sequence using a limited number of installation vessels and project-specific tools and equipment.

Nowadays, subsea systems are designed with installation flexibility as a core requirement. Over time, engineers have created flexible but reliable modular systems for the deepwater. The introduction of jumpers and flying leads reduced scheduling and interdependency conflicts during the installation cycle. *Heave compensation devices,* operated from the back of large work boats, expanded the selection of vessels available for installation tasks beyond the drilling rig. For example, traditionally, the last job of a drilling rig before it leaves the drill location was setting the tree. Now work boats are outfitted to do this job, figuratively replacing a sledge with a ball peen hammer. (*See* Fig. 8–6.)

Figure 8–7 shows multiple tasks underway simultaneously at Shell's *Crosby* project in the Gulf of Mexico. In the foreground a drilling rig is completing a well. At mid-view, an umbilical

Fig. 8–6 A
Subsea Tree
Component
Being Lowered
from the Back of
a Work Boat
(Courtesy Shell
Exploration and
Production
Company)

Fig. 8–7 Several Vessels Working Simultaneously at the Crosby Project (Courtesy Shell Exploration and Production Company)

installation vessel prepares to lay the umbilical back to the host. At the back, a pipe lay vessel gets ready to lay lines from a subsea manifold to the host. Buzzing around it is a support service vessel with an ROV to monitor the pipe lay progress.

The scheduling and implementation of these multiple efforts creates a new role in the process, the *installation coordinator*, the impresario of a safe and efficient project.

Manifolds. Because they come in many sizes and shapes, manifolds are lowered to the sea floor in a variety of ways. Drilling rig cranes have installed many, but a more economical method is usually to employ a crane vessel. The crane lifts the manifold from the deck of the transport vessel and sets it in the water. (*See* Fig. 8–8.) Sometimes the crane continues to lower the manifold, but in the deepwater, the drop requires more wire rope than a crane normally carries. In that case, the load is then transferred to a winch that lowers the manifold as it plays out supporting wire rope from a spool. In some cases, the manifold is lowered to the sea floor by controlling the buoyancy so that a very low load capacity winch with plenty of lowering cable can be used.

The vessel used to lower the manifold into place, whether drilling rig, crane vessel, or boat, locates itself using the global positioning system satellite system. (*See* chapter 1, "Positioning by Chance.") The location of the manifold—or any other item for that matter—being lowered is determined by placing acoustic transponders on the sea floor, on the manifold, and on the vessel.

Fig. 8–8 A Crane Vessel Lowering a Subsea Manifold (Courtesy FMC Technologies)

By pinging from the vessel, and using the known speed of sound in water, distances and angles between the sea floor, the manifold, and the vessel transponders are determined. Using triangulation, the location of the manifold can be followed as it is lowered. Since the manifold is not hanging directly under the vessel because of deep ocean currents, the surface vessel has to move in iterative steps to get the manifold to the desired set down spot. With a lot of patience—but remarkably little time—the manifold can be put right on target with the correct orientation. Typical specifications call for the manifold to be placed within 5 ft of the designated spot, within 5 degrees orientation, and less than 5 degrees off level.

Jumpers. These pipe segments, 50–100 ft long, with special end termination fittings, connect the flowline to the manifold or to the gathering line sled. In some architectures, the wells are clustered around the manifold, and the jumper connects the well directly to it. The metrology needed to fabricate a jumper relies on ROVs using a range of measuring devices, from simple steel tape measures to acoustic apparatus

Mechanical devices operated by an ROV measure the horizontal and vertical offsets, and associated angles between the two connecting points. Once the final measurements are made, the jumper pipe sections are fabricated to match. They are lowered with a winch system and guided into place by an ROV using its video display. To aid in orientation, one end of the jumper is painted with white stripes or some other distinguishing characteristic, the other end with different markings. The lowering winch and ROV operators then always know which end they are monitoring. Even with high-tech lights on the ROV, the viewing range may only be a few feet, certainly not the 100 ft of an entire jumper.

By coordinating the lowering action of the winch with some gentle lateral nudging from the ROV, the hubs on the jumpers are aligned with the posts on the manifold (or sled or tree) and an initial connection is made. Structural, mechanical, and flowing pressure integrity is then completed and assured as the ROV provides hydraulic pressure through a connecting port on the hub to force alignment, connectivity and pressure sealing.

Umbilicals. Shipped from the factory on large steel reels, an umbilical is installed from the back of a work boat by unwinding it from the spool and lowering it to the sea floor. Although short segments of an umbilical are stiff, over the long distance from the host platform to the manifold, an umbilical can flex into a catenary shape, or snake around obstacles without buckling or otherwise losing integrity. ROVs do the subsea connections the same way as for the jumpers. The host-end of the umbilical is transferred from the installation vessel to the platform, where the connection is made on the deck in a much dryer and accessible environment.

In many cases, only electrical or hydraulic or chemical lines need to be connected to a well or sled from the manifold, without the rest of the umbilical services. A smaller umbilical can be installed by having the ROV "fly" each end to its place, plug them in, and do the mechanical and integrity checks.

Subsea systems involve many elements, so system designers work hard to reduce the work vessel schedule interdependencies during installation and minimize the logistical effort involved in choreographing the time and space needs of the various work vessels. For example, a particular subsea system might have the following installation sequence: (a) install all subsea trees; then (b) install all flowlines; then (c) install the manifold; then (d) install gathering lines and umbilical; then (e) install jumpers and flying leads. With proper design and appropriate planning, the sequence could be modified to *c–b–a–d–e* or to *d–a–b–c–e*, or several other combinations—as long as *e* is last. This flexibility allows for the inevitable but unforeseeable schedule problems and gives the opportunity to use a wide range of vessels on an opportunity basis, but it needs cooperative innovation between designers and contractors and the oversight of the installation coordinator.

System reliability

Because the subsea elements are way down there and hard to get to, designers and builders emphasize redundancy and reliability—not unlike like the space industry. But items do fail and need replacement. Like the control pods on the tree, many systems are modularized so that they can be removed and replaced with relative ease. Though an ROV contractor is never satisfied that enough is done, system designers build in extensive ROV-friendly connections and contact points that make the ROV's intervention simple and straightforward.

The operational up-time of the average subsea systems is remarkably good, approaching—and sometimes exceeding—that of the more accessible surface facilities in many areas. Subsea well problems are usually related to reservoir issues and not hardware problems.

ROVs

How does a tree get bolted to the wellhead in 7000 ft of water? How does the crew of a lift barge find the end of a mooring line? How does anyone replace a damaged seal ring on the bottom of a blowout preventer (BOP) being put in place? And how does anyone know it's damaged anyway? The answer to all these questions, and the *sine qua non* for subsea systems, is ROVs.

Almost every deepwater drilling rig has a sidekick ROV assigned to it. In most cases, a company other than the rig operator provides and operates it. That's how specialized ROV technology is.

ROVs in E&P service fall into two general categories, *inspection* and *work class*. *Inspection* ROVs are smaller, cheaper, and provide only a set of eyes in the submerged depths. *Work class* ROVs can carry some 400 pounds of payload—hydraulic and electro-mechanical tools and supporting supplies. (*See* Fig. 8–9.)

Inspection class ROVs can do their surveillance on underwater structures and pipelines by camera on the fly. They can also stop and connect their sensors to check cathodic protection or structural integrity. When connected to pipelines, they can check wall thickness and listen for sand particles and other plugging problems.

The work class ROV, a subsea handyman does more diverse tasks. In fact, a work class ROV can have its own tool rack nearby where it stows wrenches, torque tools, cable cutters, awls, pincers, and other devices. *Manipulators* (articulated arms) will pick them up and drop them off as the job demands. Manipulators typically have the seven degrees of movement—turning, elbowing, hinging, grabbing, rotating, etc.—that the human arm has.

Fig. 8–9 Work Class ROV (Courtesy Oceaneering)

The supplies in the payload can include, for example, a cargo of hydraulic fluid that the ROV delivers into a waiting tank in seabed equipment that has an ambient pressure of 10,000 psi or more.

An ROV has lights and a television camera that relays a reasonable image back to an operator's console on board the rig or platform. The image provides the operator a 2D view of a 3D underwater work environment—challenging, but adequate. For some jobs, a work class ROV brings along a buddy, an inspection ROV. That provides a camera view form another angle—not quite 3D but useful for obscured or especially touchy approaches.

Maneuvering. Propellers driven by electric motors or by electric-activated hydraulic pumps provide propulsion. The 75–100 horsepower motors deliver about a 1000-pound thrust. The ROVs are nearly neutrally buoyant, so most ROVs do their vertical traveling to and from the target depths in cages suspended by umbilicals (Fig. 8–10) to allow rapid transit and minimal deviation from a straight line that currents might cause. At the target depth, the ROV exits the cage, to which it remains connected by a tether. It can roam the length of its tether, some up to 3000 ft. On return, the ROV uses its tether, sonar, and a transponder on the cage to find its way home. The operator reels in the tether, and the ROV pops back in the cage like a well-trained Labrador retriever.

Wonderful as that sounds, ROVs have limited dexterity. The problem arises when they have no stable platform. Doing precise work with an ROV is analogous to threading an earthbound needle from a hovering helicopter. For that reason, connections between an ROV and a subsea fixture are designed as crudely as possible to accommodate the underdeveloped gross motor skills of the ROV. Most of the time, an ROV has to lock on to a subsea fixture (a wellhead, a manifold, a tree) before it can switch to its finer motor skills. The ROV can use one of the manipulators to grab the structure. Alternatively, if the engineers

Fig. 8–10 ROV in Its Cage Being Launched (Courtesy Oceaneering)

who designed the target equipment have done it right, the ROV can lock into a waiting receptacle and immediately start turning valves or doing whatever the task might be. Figure 8–11 shows the sequence of tasks as an ROV attaches a pipeline to a subsea manifold.

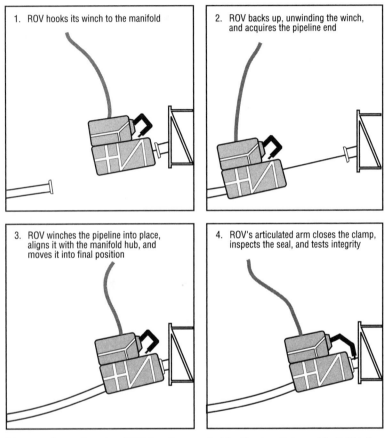

1. ROV hooks its winch to the manifold

2. ROV backs up, unwinding the winch, and acquires the pipeline end

3. ROV winches the pipeline into place, aligns it with the manifold hub, and moves it into final position

4. ROV's articulated arm closes the clamp, inspects the seal, and tests integrity

Fig. 8–11 Sequence of an ROV Attachment of a Pipeline to a Subsea Manifold (after Oceaneering)

Umbilicals and tethers. The ROV connects to the cage through a tether; the cage connects to the controller on the surface through an umbilical. Wires run through both to deliver electrical power. All the signals to and from the ROV are digitized, as are the camera images, and all travel through fiber optic cables in the tether and umbilical. Hydraulic fluids that run ROV devices don't move via the umbilical. They are stored in hydraulic reservoirs on the ROV. Electric motors onboard turn the hydraulic pumps.

Besides doing transmission, the umbilical also acts as the winched line to lower and raise the ROV and cage. The steel *armor shield*, the umbilical cover, bears the weight. As the ROV moves to greater depths, the weight of the umbilical is self-limiting. At 10,000 ft, the umbilical approaches the limit at which the armor shield can hold up its own weight with acceptable assurance. At 15,000 ft, Kevlar, the same polymeric material used for personal armor, is substituted for steel. It weighs less and needs less strength for the same length. It is, however, more than four times as expensive as steel and requires more careful handling.

Operators. The operator maneuvers the vehicle with a joystick, watching screens on his console that relay information from sonar and the camera on the ROV as well as the transponders on the seabed or equipment being serviced. The sonic guidance system has an inherent error of plus or minus a half percent of the water depth, so the operator has to rely on both the sonar/transponders and the television images. ROV operators say that getting the vehicle to the right spot is akin to landing a helicopter underwater at night. As pilots do, they train on simulators (Fig. 8–12) to establish their proficiency.

Most of the time, ROV operators work under air traffic controller-type stress. First, the subsea environment is filled with physical obstacles to success: equipment, flowlines, jumpers, flying leads, and even the ROV tether. Second, ROVs tend to be painfully slow anyway because of the sluggish environment they work in, the limited visibility, and because they are—well, remotely operated. Finally, the operators realize that every moment they spend doing their specialized work—connecting, fixing, adjusting, or whatever—the cash register of an idled

Fig. 8–12 ROV Training Simulator (Courtesy Oceaneering)

$100,000 to $200,000 per day drilling rig is ringing. An ROV manager will tell you that the ideal operator is a former Nintendo adolescent who grew up to be a helicopter maintenance technician and who scuba dives as a hobby.

9

Topsides

Crowded to the full with glorious action,

and filled with noble risks.

From *Count Robert of Paris*

Sir Walter Scott, 1771–1832

The name says it all—the place atop the platform where the drilling and processing equipment sit, where scores of auxiliaries continuously labor away, and where the crew makes its temporary home. Topsides for FPSOs may look remarkably different from other floating systems and fixed platforms, but the list of necessities remains the same.

Production engineers wish that oil and gas would come out of the ground ready for sale, but most reservoirs disappoint them. In almost every case, a cocktail of oil, gas, water, and solids come up through the riser to the platform. The mixture could be predominantly oil with some gas or the reverse. In either case, water usually accompanies the hydrocarbon. (*See* Fig. 9–1.)

Oil/gas/water mixtures are easy enough to deal with, but each of them can bring a host of bad actors that demand attention. Dissolved and solid minerals, especially salt and inorganic scales,

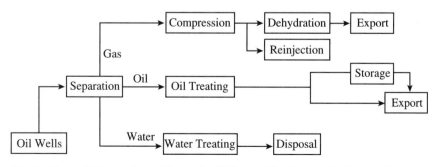

Fig. 9–1 Crude Oil Processing on Deepwater Platforms (after Paragon Engineering)

Tarnish

The faintly oriental terms, sweet and sour, refer to the amount of sulfur contained in either crude oil or natural gas. During the earliest commercial oilfield operations in Pennsylvania in the 1860s, wildcatters searched for oil to satisfy the growing demand for kerosene. The shortage of whales led to a decline in the whale oil used for residential and commercial lighting. Kerosene became the easiest substitute.

As this industry grew, many people noted that burning certain grades of kerosene emitted rather bad odors and caused their silverware to tarnish rapidly. The root cause, they came to find, was the sulfur content. At the same time, oil producers found they could differentiate high and low sulfur content kerosene by their tastes—high sulfur kerosene tasted sour, low tasted sweet. For years tasters rated kerosene batches, always searching out the sweet-tasting, premium-priced grades.

Eventually more analytical methods replaced the hapless tasters, but the unsavory nomenclature persists to this day for both oil and gas.

come along with the water. The oil can carry sand and dirt from the reservoir and scale and corrosion products from the tubulars.

Fortunately, through some geological quirk, most offshore natural gas has minimal content of carbon dioxide (CO_2), hydrogen sulfide (H_2S), and other *acid gases* common to many onshore areas. With these small concentrations, expensive apparatus to remove them is not needed. However, if much water vapor is present in the natural gas, even the small amounts of acid gas can dissolve in it and create corrosive compounds that attack topsides equipment and raise havoc with pipelines. In that case, chemicals are added to inhibit corrosion before processing or transportation. Some cases even call for processing equipment fabricated from special, corrosion-resistant alloy steel.

Deepwater crude oils are generally sweet, that is, most have less than 1½% sulfur. Even for those that contain more, the sulfur does not present the potential corrosivity that the acid gases in natural gas do. Sulfur in oil comes in the form of various complex hydrocarbon compounds that usually do not readily react to form corrosive compounds. So the oil goes through no treatment on the platform to sweeten it, leaving the refineries to handle that problem.

To deliver marketable products into the export riser, the oil, gas, and water must be separated from each other and the unwanted contaminants and trash removed. Then the water and leftover trash must find an environmentally acceptable home.

Marketable oil, a loose specification negotiated with the pipeline or tanker company taking it away, generally needs a *basic sediment and water* (BS&W) content of no more than 0.5–3.0%. This BS&W can be "salty," having a salt content several times higher than ordinary seawater. Often there is an additional requirement that the salt content of the oil cannot exceed 10–25 lbs per thousand barrels. Facilities at the receiving refineries clean up the rest.

Oil Treatment

Every platform is designed to accommodate the unique combination of fluids expected to be coming up the riser, including even anticipated future production from other fields and subsea developments. Of course the designers don't always get it right. Newly discovered fields nearby might be tied in; production from new zones in existing fields might have different characteristics. That might require retrofitting topsides facilities to handle different volumes and compositions over the lifetime of a platform.

Separation. For this section, think of the hydrocarbons coming from the riser as mostly oil with some associated gas. The first block in Figure 9–1 involves separating the gas from the liquids. The well stream from the riser flows into a separator, a cylindrical pressure vessel, sometimes vertical, sometimes on its side. The sudden increase in volume and decrease in pressure causes the "beer bottle" effect. Just as the carbonation starts to leave the beer when the top is popped, the natural gas *flashes* or separates from the oil. Gravity pushes the liquid to the bottom, where it is drawn off, leaving the gas at the top. Everyone knows oil and water don't mix, and conveniently the water—or most of it—drops to the bottom of the separator, leaving the oil in the middle. The three separated phases leave the separator through separate nozzles, heading for the next phase of treatment.

Actually, oil/gas separation often takes place in several stages. (*See* Fig. 9–2.) Even with the sudden drop in pressure, and even if the associated gas comprises only a small percent of the total, some gas may remain dissolved in the oil. Oil flows from the first stage, a high-pressure separation, to a second stage and third stage, where pressure reductions cause more gas to flash. At the first two initial stages, any entrained water follows the oil path. In the last stage, the water is drawn off the bottom of the separator. By that stage, *live oil* has transformed into *dead oil*. For arcane reasons having to do with vapor/liquid phase equilibrium, reducing the pressure of the well stream fluids

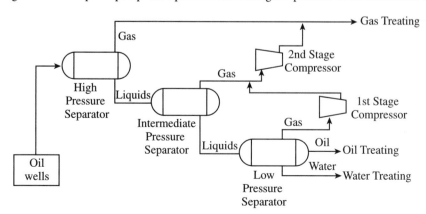

Fig. 9–2 Multistage Gas Separation on Deepwater Platforms (after Paragon Engineering)

in multiple stages results in a better gas/oil split than a single separator would and results in dead oil with a lower vapor pressure, which is good for transportation purposes.

The first stage, where most of the gas comes off, runs at a pressure higher than the export riser requirement. That way, most of the gas, even with pressure drops through downstream treating, can flow with no boosting to an export line. Reinjection of the gas would require additional compression. The gas coming from the latter stages has to go through compressors to reach the pressure necessary to commingle it with the first stage gas.

Treaters. The settling-and-flashing technique of a separator, rudimentary as it is, may not quite do the water/oil split, especially for very viscous crude oils. If the water content remains too high, the oil is sent to a water removal treater. (*See* Fig. 9–1.) Many different designs are available. The treaters lower the viscosity or break oil/water emulsions by heating, electric current, agitation, or other means to increase the droplet size of the water contained in the crude oil. This in turn is followed by quiescent sections where the water settles to the bottom.

From the treaters, the dry oil—or as dry as it has to be, given the BS&W spec—flows to a rundown tank, where a pump moves oil continuously into the export riser or to a larger tank for transport in large batches to a shuttle tanker.

Most streams arriving at most deepwater platforms have only moderate amounts of solids in them, but where they occur they have to be separated using sedimentation settlement tanks, hydrocyclones, and filters. The solids can then be cleaned sufficiently by washing with relatively oil-free produced water and disposed of overboard or hauled from the platform in workboats to onshore disposal sites.

Water Treatment

Just as the oil from the separators still contains water, water from the separator still contains some oil. The logical destination of the water is overboard, but its quality and environmental concerns may call for reinjection into the subsurface. In either case, the oil needs to come out of the water, and solids removal may be required also.

One of several types of processes, *skimmers* use a long residence time in a vessel with a large surface area to allow the small droplets of oil to rise. The top layer, heavy with oil, passes over a weir or is skimmed mechanically. Sometimes stubborn oil emulsions that hold in the water call for chemical additives to speed the separation process.

Plate coalescers, an alternative separation process, use closely spaced plates. These close plates reduce the distance that the small oil droplets must rise before being captured by the plate and channeled to the top for skimming.

Another innovative device is the *hydrocyclone*. The water/oil mix is squirted into an inside of a cone (apex up) at high velocity via a tangential inlet. This causes the fluid to "spin" at a high rate of speed in a cyclonic manner, forcing the heavier water down and out along the inside surface of the cone and to exit the bottom. The lighter oil moves upward and inward to the center axis of the cone and out the top. Often the topsides have two more of these processes in series to progressively reduce the oil content of the water.

During the oil/water separation processes, some natural gas comes off. A line from the top of the vessels moves it to a convenient injection point in the gas treatment section.

Gas Treatment

Heating. A typical gas well has several thousand pounds per square inch of pressure at the wellhead, much higher than needed to treat it on the platform and export it. As the gas moves through the *choke*, a carefully designed constriction in the flowline, the pressure drops, the volume expands, and following *Boyle's Law*, the temperature drops precipitously. Residential and commercial air conditioners depend on this same principle: the temperature of an expanding gas falls.

As the temperature drops, whether the wellhead is located on the seabed or at topsides, the water in the gas stream can turn to liquid, or worse, form *hydrates*. Snow-like in appearance under close scrutiny, hydrates form as methane trapped in icy crystals. They can easily plug a flowline. Hydrates can form at any temperature between 30 and 80° F even though water by itself freezes at the low end of the range, 32.2° F.

When hydrates plug a wellhead or flowline, defrosting them can be difficult and perilous. Hoping that once the flow stops, the hydrates liquefy is akin to thawing frozen water pipes in Alaska: wait until spring—maybe. And when the hydrates do melt and release the backed-up pressure, equipment downstream can get a sudden and dangerous jar.

To avoid hydrate formation and the plugging that goes along with it, the chokes and a segment of the flowlines from a topsides well head are enclosed in a heated water bath, kept at a high enough temperature so that the gas is always above the hydrate-forming temperature. (*See* Fig. 9–3.)

Separation. Even if the gas coming from the riser originated in a gas well, it runs through a separator to remove *condensate* and water in the same way described for oil. Water drops out the bottom, oil from the middle, and gas from the top. Though not shown in Figure 9–3, the separation

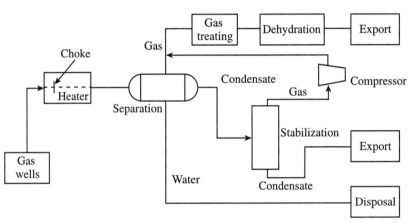

Fig. 9–3 Natural Gas Processing on Deepwater Platforms (after Paragon Engineering)

can be run in the three stages of Figure 9–2. Condensate has the composition and other characteristics of very light crude oil.

Stabilization. Sometimes the multistage separation process is not sufficient to stabilize the condensate. The oil/condensate coming out of this simple piece of equipment could still have more-than-desirable volatile gases dissolved in it. Since the condensate gets stored in a tank aboard the topsides, at least momentarily and sometimes for a day or more, the volatility presents a hazardous condition. The *vapor pressure* of the condensate, a measure of the content of light gases such as butane, propane, ethane, and even methane, can be reduced by passing the condensate through a *stabilizer.*

The condensate first goes through a heater and then flows into the middle of a column, maybe 30 ft tall, with perforated trays or packing inside. The light gases work their way to the top, wringing out any liquids, which fall toward the bottom. As the liquids fall to the bottom, the sloshing action through the trays or packing agitates any volatiles out and towards the top. Because of these two actions (wringing out and agitation) on each tray or level of packing, additional separation occurs.

The gas from this step goes through a compressor and then commingles with the high-pressure gas stream for treatment and dehydration.

Gas dehydration. Gas going into the export riser has to be dry. That is, it has to meet maximum water content. Otherwise, the water could condense and accumulate in low spots, form hydrates, or otherwise constrain flow. Also, water combined with small amounts of the acid gases forms a slow but corrosive reaction with almost everything it touches, pipelines, treaters, compressors, and all the other equipment aboard.

A *glycol dehydration system* is most often used to remove the water from the gas stream. The gas flows into the bottom of a contact column with trays or packing. (*See* Fig. 9–4.) Triethylene glycol (TEG) enters the top of the column. TEG has an affinity for water, and as it sloshes past the gas, it absorbs most of the water and carries it out the bottom of the column. Dehydrated gas exits the top. Dehydrated gas is not to be confused with *dry gas*, which is natural gas with the *natural gas liquids* (propane, butane, and natural gasoline) removed.

The TEG then goes to a regeneration step. First it goes to a separator, where any entrained oil can be skimmed off. Then it goes to a heater where the temperature is raised to about 400° F. That causes the water to vaporize, leaving behind the TEG, whose boiling temperature is 549° F. The regenerated TEG goes back to the top of the contact column for reuse. The water vapor goes either to the atmosphere or to a cooler to condense and capture any hydrocarbons that made it through the whole process.

In an alternative dehydration process, a *desiccant* or *absorbent* like silica or alumina gel in two parallel steel tanks captures the water. The gas stream flow is alternated between the tanks: while the absorbent in one tank is heated to drive the collected water off to the atmosphere, gas passes through the other tank to have the water removed.

Some topsides have refrigeration or membrane permeation systems to do the dehydration. Once the gas is properly treated, it is sent into the export riser, on its way to market. In most deepwater environments, the seafloor temperature runs at 30 to 35° F. Since

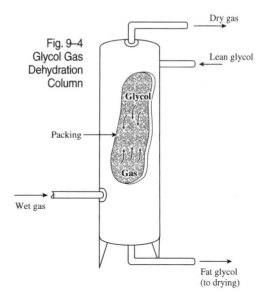

Fig. 9–4
Glycol Gas
Dehydration
Column

Dry gas

Lean glycol

Glycol

Packing

Gas

Wet gas

Fat glycol
(to drying)

the gas export riser and the pipeline run along the sea floor, the temperature of the gas soon drops to the seafloor temperature. Typically the water content (in vapor form) is limited to three or four pounds per million cu ft of gas. At the normal operating pressure of the export line (1000 to 1100 psi), with that much water, the gas has a dew point of about 0° F. (Dew point is the temperature at which water begins to condense in the form of droplets. Dew point varies with pressure.) That gives a 30–35° F margin to keep the water in the vapor phase.

Metering

Sending oil and gas into the export risers can be like pumping money overboard if accurate metering doesn't take place. For oil, a *lease automatic custody transfer (LACT)* unit measures volume. LACT units come in many sizes and levels of sophistication. The larger units have built-in *meter provers*, devices that automatically and periodically recalibrate the meter. Most LACT units also have probes to monitor BS&W content. Some even block the flow to the meter if the crude oil fails to meet the BS&W specification. After all, who wants to pay $25 per barrel for water and sediment?

The mechanism in a LACT unit is most often a positive displacement (PD) meter, a direct measurement device. It divides the flow into discrete packets and counts the number passing through the meter. The PD operates by having the flowing oil drive its internals of vanes, pistons, or rotors, which displaces a known and specific amount for each revolution. The revolutions are counted, either by a mechanical device, or more commonly by reading electrical pulses, measuring the total volume. Readout devices provide a continuous update of status.

Another type, the turbine meter, measures volume flow indirectly. It belongs to the "inferential" family of meters. Turbine meters infer the volume by measuring the rotation speed of a turbine suspended in the flow stream. PD meters are more widely used for custody transfer, but if they are properly calibrated and maintained, turbine meters can also be used.

Another form, the ultrasonic meter, uses the transit time principle. Signals emitted from transducers travel faster when moving *with* the flow than when moving *against* the flow. The difference in transit time is used to calculate the flow rates. Ultrasonic meters have found some use in gas pipeline applications. As more proprietary research on the ultrasonics improves accuracy, reliability, and cost, they increase their penetration of the measurement market.

For gas, *orifice meters* are used most often to measure the flow. An orifice meter has three essential elements:

(1) a conduit or pipe through which the gas flows; the conduit contains

(2) an orifice plate, which is a flat steel plate with a precise hole in the middle

(3) a means to measure the difference in pressure caused by the disturbance created by the orifice plate, from its upstream side to its downstream side

This differential pressure is proportional to the square of the flow velocity, and so it follows that the flow rate can be calculated from the pressure differential. Orifice meters have a long history dating back to the Italian physicist Giovanni Venturi. They are reliable and accurate, with a range of plus and minus 0.05 %, a usually acceptable tolerance.

Personnel and Their Quarters

Complement. The crew size on an offshore facility depends on the number and complexity of wells, the equipment, and overall philosophy of the operating company. Some platforms have advanced electronic and computer-assisted controls that run much of the operation; others are more reliant on people to operate. As an example, say a certain facility needs a full-time operating crew of 60 to operate and maintain—40 during the daytime, and 20 at night. (Some jobs only require a daylight shift.) Another 60 are on their days off and about 15 more have to be available for vacation coverage, sick leave, and other contingencies. All in all, the complement adds up to 135 people, about 2¼ times the daily contingent.

Shifts. For most offshore operations, the crews stay onboard and operate the facility for one week, working 12-hour shifts, then have one week off. In the deepwater, the crews may do two weeks on and two weeks off.

Quarters. All these qualified people need a place to spend their off-duty time on the platform. They need sleeping accommodations, food service, fresh water, electricity, laundry, stores, medical facilities, relaxation rooms, and workout rooms. Figure 9–5 shows the dining features in the crew's quarters on the *Ram Powell* TLP in the Gulf of Mexico. For obvious safety reasons, personnel facilities sit as far from the drilling, the wellheads, and the hydrocarbons handling equipment as possible. No one questions that the design, comfort, and safety features of these facilities lends itself to the efficiency and effectiveness of the operation.

The galley on a production platform typically serves four meals daily—the normal three plus a midnight meal for the night crew. Snacks are set out often during the day. In a typical week during peak times, the *Ram Powell* galley serves 2600 meals, using 160 dozen eggs, 1350 lbs of meat, 500 lbs of potatoes, and the crew does 840 loads of laundry.

Fig. 9–5 Crew Quarters on a Deepwater Platform (Courtesy Shell Exploration and Production Company)

Personnel transportation. For most deepwater platforms, crews move to and from shore in large helicopters, sometimes carrying as many as 24 passengers. Even so, large facilities need several round trips during a shift change. A helicopter-landing platform sits on the top deck, generally above the crews' quarters, because it too needs to be as far from the operating equipment as possible for safety reasons. Some deepwater operations are close enough to shore bases and have calm enough waters that personnel can move in cheaper crew boats. Crews transfer from the boat to the platform in a transfer apparatus know as a "Billy Pugh basket," lifted by the platform crane.

Safety Systems

The safety of people working around complex equipment, in an environment that is itself quite volatile, demands the full attention of the system designers and the actual operations staff as well.

If process controls always worked, if equipment always performed perfectly, and if personnel were always careful, safety would not have to be systemic. But they don't, so it is. The design stage for equipment and facilities integrates safety with operations. In practice, that calls for measuring pressure and temperature and other operating parameters at many places on the platform, with feedback loops to allow corrections or shutdowns if the process gets beyond safe limits. It also means fire retardant material covering structural members, wash stations where they might be needed, hazard detection systems, sprinkling systems, and other damage control apparatus.

Operating procedures call for periodically maintaining, testing, and calibrating the safety devices. Industry organizations such as the American Petroleum Institute, the Norwegian Oil and Gas Partners, United Kingdom Offshore Operators Association, and the International Standards Organization publish guidelines for all aspects of safety. In the end, the most important safety feature is pervasive safety awareness.

During the design phase, Hazardous Operation Reviews evaluate a thorough list of "What if" questions. Suppose a certain valve malfunctions. How is a ruinous failure averted? What if a pressure monitor fails? What's the back-up? What if a pressure relief valve pops? And so on.

Also during design, while the crew goes through operations training, they learn how to watch for the development stages of dangerous conditions and how to react to a full array of calamities. The money spent on attention to safety during the design phase easily reaches 50% of the topsides costs.

Large platforms have a last escape route—crew capsules. Designs vary, but they are the equivalent to lifeboats on a passenger ship. If a serious failure occurs—explosion, uncontainable fire, or the failure of the platform—the crew flees to the capsule and leaves the platform. Capsules sit in several strategic places around the platform, with access paths clear at all times as part of the safety program. Figure 9–6 shows a 40-person escape capsule. Normally a platform has enough capsule capacity to handle one and one-half times the onboard complement.

Fig. 9–6 Survival Capsule aboard a TLP (Courtesy Survival Systems International)

Auxiliary Systems

Extensive (and expensive) as the main functions just summarized are, the auxiliary systems supporting them often require more engineering time and effort, and capital expenditures. Figure 9–7 shows an overall layout of topsides on a fixed platform. Figure 9–8 shows an exploded view of the *Mars* TLP topsides, which was installed in large modules, as seen in Figure 9–9.

Power. Most equipment on board is powered by electricity—pumps, lighting, personnel facilities, ice cream machines, and computers. The odd gas-fired or diesel engine shows up occasionally on topsides, but the design, installation, maintenance, and operation of electricity-driven equipment makes most sense.

For platforms with an adequate supply of natural gas, gas-fired electric generators are always chosen. The next best alternative, a diesel-fired generator, requires the diesel fuel be transported to the platform and stored onboard. Even then, turning diesel fuel into electricity to drive rotating equipment, such as compressors and pumps, makes more sense than placing direct drive gas-fired or diesel engines around the platform. For that reason, gas-fired or diesel engines rarely find a home on board except for electricity generation and for large compressors (greater than 1000 horsepower) and pumps.

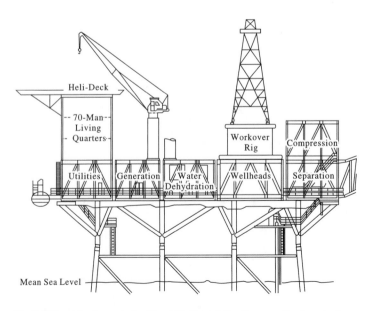

Fig. 9–7 Topsides Layout of a Fixed Platform (after Paragon Engineering)

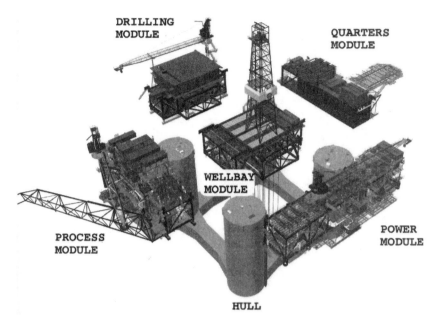

Fig. 9–8 Exploded View of *Mars* TLP Topsides (Courtesy of Shell Exploration and Production Company)

During peak activity levels, electricity consumption can chew up large volumes of gas or diesel. The *Ram Powell* TLP in the Gulf of Mexico, during its drilling and completion phase, generated enough electricity to have supplied 7000 homes. The various turbines and engines around the *Ram Powell* topsides have the equivalent power capabilities of about 225 big Ford pick-up trucks, the Model 350 V8s.

Pumps. The ubiquitous piece of equipment on the topsides is the pump. Big pumps move oil into the LACT unit; water to reinjection or disposal; oil to treating. Smaller pumps move skimmed oil, methanol, cooling water, firewater, and so on.

Cranes. Most of the equipment, maintenance supplies, and sustenance needs arrive at a deepwater facility by supply boat. Platforms have at least one crane, and some two or three, to offload their supplies.

Flares. Almost all regulatory jurisdictions require natural gas flaring on a continuous basis be kept to a minimum. Still, safety and engineering design call for an outlet for the natural gas, especially for emergency conditions since it cannot be stored in a surge tank like oil.

Fig. 9–9 Deck Module Being Lifted onto the *Mars* TLP (Courtesy of Shell Exploration and Production Company)

Piping from the affected equipment is run across the topsides and goes up a riser into a flare tower or a flare boom. That takes the flare flame away from both the people and the facilities. A flare has both an ignition system and a constantly lit pilot to ensure instantaneous response to emergencies.

Other auxiliaries. In and around the recesses of the topsides are more facilities:

- control rooms
- boilers for steam for heat and power
- electricity distribution and switch gear rooms
- instrument air and power
- fresh potable water systems
- sewage treatment
- boiler feed water and reinjection water treatment
- fuel gas treatment

Export connections. The fluids that flow into the platform topsides may sound like a witch's brew—oil, gas, water, acid gases, sand, and more. The separation and treatment facilities transform them into marketable quality hydrocarbons that flow into the export risers. Chapter 10 continues the story with a look at risers, pipelines, and flowlines.

10

Pipelines, Flowlines, and Risers

Oh me, why have they not buried me

Deep enough

From *Maud*

Alfred, Lord Tennyson (1809–1892)

"Nothing happens until we get our lines in place," say the subsea pipeline engineers, and they are right. Oil and gas have to get from the reservoir to a refinery or a gas transmission system to make investments in the well, the platform, and the topsides pay off.

The logical way to deal with this subject is to follow Figure 10–1. The fluids flow from the wellhead through a *jumper* to a manifold. From the manifold, the admixed fluids move through a

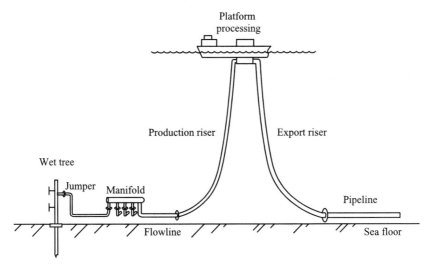

Fig. 10–1 Well Production Flows from the Wellhead to Shoreline Connections

flowline, which could be up to 70 miles or longer for gas and 10–20 miles for oil, to a *production riser*. That line moves the fluids to a production platform where processing takes place. Outbound from the platform, the fluids move through an *export riser* to a *subsea pipeline* and on to shore.

While they have different names, flowlines and pipelines differ from each other mostly in size. Other characteristics and installation techniques are similar. Jumpers and risers have their own issues.

Line design

The further downstream from the wellhead, as more streams commingle, the larger the diameter. Lines, of course, are sized to handle the expected pressure and fluid flow. Jumper diameters can be as small as 4 in., but 6- to 12-in. pipes are common. Pipelines in the deepwater generally range from 12 to 30 in. To contain the pressures, wall thicknesses of the pipe range from ½ in. to 1¾ in. of steel.

The temperature of water in the deep ocean hovers around 30–35 °F. The water rapidly sucks heat out of the fluids in a bare steel pipe. The freezing cold environment can cause hydrates to form in a gas line or paraffins, waxy hydrocarbons, to plate the walls of an oil line. Either way, the line can plug, making a very bad day for the production crew on the platform, some 1500 to 10,000 ft above. To avoid this disaster, most flowlines are coated with an insulating material to keep the well fluids warm. (Injection of antiplugging chemicals at the wellhead can provide a complementary antidote to plugging.)

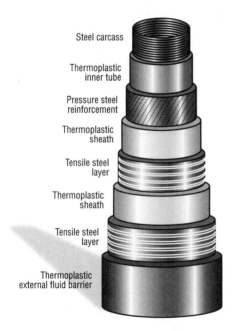

Steel carcass

Thermoplastic inner tube

Pressure steel reinforcement

Thermoplastic sheath

Tensile steel layer

Thermoplastic sheath

Tensile steel layer

Thermoplastic external fluid barrier

Fig. 10–2 Flexible Pipe Construction (After Welkstream)

Oil and gas export pipelines are usually not insulated. The topsides treating facilities dehydrate the gas to remove the hydrate-causing water vapors. Oil pipelines are periodically cleaned to remove wax or paraffin build-up in the pipe walls.

As an alternative to externally applied insulation, *pipe-in-pipe* works very well. For this technique, a 12-in. line is enclosed in an 18-in. line, for example. The annular space between the two lines is filled with a high-quality insulating material, providing a thermos bottle-like effect. Insulated lines sometimes get roughed up during installation, tearing the insulation, exposing them to cold spots and possibly causing plugging problems. The pipe-in-pipe eliminates this hazard.

Although the lion's share of pipelines and flowlines are steel pipe, there is an alternative, flexible pipe, fabricated from helically wound metallic wires interspersed with thermoplastic layers. Each layer has a specific function. (*See* Fig 10–2.) These *flexibles*, as they are called, have excellent bending characteristics and are often used where steel pipe would be too rigid.

The Boon and Bane of Buoyancy

During the installation of a pipeline, as the pipe reaches from the stern of a *lay barge* to the seabed, the entire vertical load of the pipe weighs on the barge. Take as an illustration, a 24-in. steel pipeline with a 1-in. thick wall. One foot of that line weighs about 250 lbs on the barge deck. In the water, it weighs only 50 lbs—if it is filled with air and not water. The buoyancy effect provides the counteracting upward 200-lb force.

Suppose a barge was laying that pipe in 3000 ft of water. It has to support more than 150,000 pounds of weight off its stern—actually more, because the trajectory from the lay barge to the touchdown point of the pipe is more than the water depth. (*See* Fig. 10–3.) If the line suddenly flooded with water because it buckled or some other nightmare occurred, the load on the system would jump to more than 750,000 pounds. That sudden increase could cause the loss of the pipeline or damage to the lay barge. Worse, as the water depth increases towards 10,000 ft, the barge would be in peril if it had no quick disconnect capability. For that reason, the laying

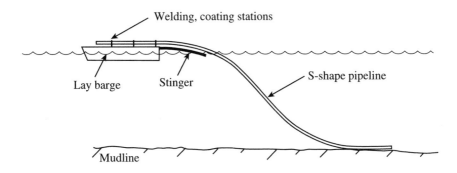

Fig. 10–3 S-lay Method for Deepwater Pipelines

procedure includes careful and continuous monitoring of the configuration of the outstretched pipeline to ensure structural integrity.

Ironically, once on the sea floor, a pipeline needs more net *downward* force to keep it from drifting. Oil, so much heavier than air, adds to the weight of the steel pipe. Gas, however, doesn't provide the extra weight that oil does. Depending on the line size, extra ballasting might be needed to hold the line in place. In shallow water, the most cost-efficient way to add weight is coating the pipe with concrete. In deeper water, the pipe wall thickness required to resist hydrostatic pressure is often enough to provide the needed weight.

Laying Pipe

Jumpers, being short and typically weighing 10–20 tons, are generally prefabricated and lowered into place using cranes or winches on the deck of the construction vessel and assisted by ROVs. In addition, ROVs provide reconnaissance for selecting routes, surveillance during the laying, and aid in connecting various pieces when they are in place, as described in chapter 8.

Flowlines and pipelines extend for miles and require special pipeline laying vessels. Laying miles of pipe in thousands of feet of water starts by welding together lengths of 40–240 ft. Contrary to casual perceptions, a long section of steel pipe is quite flexible. While a 40-ft length may seem totally rigid, an unsupported 24-in., 5000-ft length droops like a fly rod with a 10-lb trout on the hook. Even more remarkable, and continuing the angler's analogy, the same pipe can be wound around a reel for transport and later un-wound during installation. To take advantage of flexibility and buoyancy, pipeline laying comes in four self-descriptive configurations, S-lay, J-lay, reel barge, and tow-in.

S-lay

An S-lay vessel has on its deck several welding stations where the crew welds together 40 or 80 ft lengths of insulated pipe in a dry environment away from wind and rain. The pipe, either for a flowline or a pipeline, is eased off the stern as the vessel moves forward. It curves downward through the water as it leaves until it reaches the *touchdown point*. After touchdown, as more pipe is played out, the pipe assumes the nominal S-shape shown in Figure 10–3.

To control the pipe curvature as it leaves the barge, a *stinger*, a long steel structure attached to the stern of the barge, supports the pipe. Some stingers extend out 300 ft. Some barges have

Fig. 10–4 S-lay Vessel, *Solitaire* (Courtesy Allseas)

articulated stingers with controllable hinges that change the shape of the stinger, managing the trajectory of the pipeline. That gives those S-lay barge vessels flexibility to operate in a variety of water depths, shallow to deep.

The double curvature S-shape requires careful control of the lay barge position relative to the touchdown point. The barge must hold the pipe at the right level of tension, otherwise the curvature can become severe enough for the pipeline to buckle, a disastrous event. Tensioning rollers and controlled forward thrust provide the appropriate tensile load.

The assembly line process, feeding, welding, coating, and laying is supported with loads of pipe arriving on transport barges as the work progresses. S-lay barges can lay as much as four miles per day of pipeline. S-lay is good to at least 6500 ft of water.

Figure 10–4 shows a world class S-lay vessel getting ready for a job. This vessel, the *Solitaire*, owned by Allseas, is 300 meters long, excluding the stinger, and can accommodate 420 people. It has two pipe transfer cranes, two double jointing plants where 40-ft pieces are joined, seven welding stations, one quality control station and two coating stations. It has the capacity to hold up to 580 tons tension on the pipeline and to lay pipe from 2 to 60 in. It can carry an inventory of 15,000 tons of pipe, which allows it to lay miles of pipeline before being restocked.

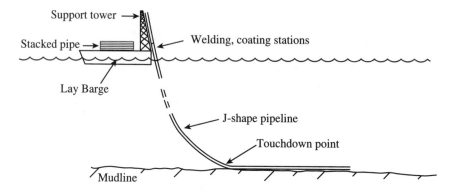

Fig. 10–5 J-lay Barge Method for Deepwater Pipelines

J-lay

To avoid some of the difficulties of S-laying—tensile load, forward thrust, and double curvature—J-lay barges drop the pipe down almost vertically until it reaches touchdown. After that, the pipe assumes the nominal J-shape in Figure 10–5. J-lay barges have a tall tower on the stern or over the side where lengths of pipe are stacked, joined, welded, and coated before they slip into the sea. The tall towers mean that realistically, only one welding station can be used, so the pipe arrives on transport barges in long, pre-welded lengths up to 240 ft.

With the simpler pipeline shape, J-lay can be used in deeper water than S-lay. During the lay procedure, the pipeline can tolerate more motion from barge movements and underwater currents that could lead to buckling. Fig 10–6 shows a J-lay system working from a large crane vessel.

Fig. 10–6 J-lay Tower on
the Crane Vessel *Balder*
(Courtesy Heerema
Marine Contractors)

Reel barge

Once they realized they could wind pipe around a reel like fishing line, pipeliners began to transport it and deliver over the stern of lay vessels that way. Some reels unwind horizontally, some vertically. (*See* Fig. 10–7.) The horizontal reels lay pipe with an S-lay configuration. Vertical reels most commonly do J-lay, but can also S-lay.

Most of the welding and coating for reeled pipe takes place onshore where costs are much less. The onshore operation requires long sections of property adjoining good dockage, as seen in Figure 10–8. There the reel barge prepares to load up with the pre-welded pipe stretched out onshore. Figure 10–9 shows the Global Industries' reel barge, *Hercules*, laying pipe in the Gulf of Mexico.

The length of pipe that a reel can handle depends on the pipe diameter. More than 30,000 ft of 6-in. pipe, the type used for some flowline applications, can fit

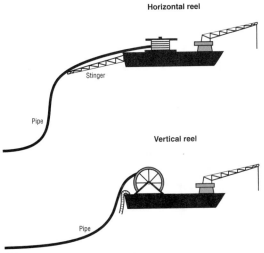

Fig. 10–7 Vertical and Horizontal Reel Barge Methods

Fig. 10–8 Reel Barge at a Yard (Courtesy Global Contractors)

Fig. 10–9 Reel Barge *Hercules* Laying Pipe (Courtesy Global
Contractors)

on a reel. In many cases, that is
sufficient for a barge to lay several
flowlines. Pipe up to a diameter of 18
in. are successfully reeled, but at much
smaller lengths. Some lay barges have
cranes that can lift loaded reels off
transport barges and return the
empties. Others have to make the
round trip to shore to do the
exchange, a time-consuming
interruption if the pipeline or flowline
are not near the supply depot.

Tow-in

There are four basic variations
of the towed pipeline method, as
shown in Figure 10–10. For the
surface tow approach, the pipeline
has some buoyancy modules added
so that it floats at the surface.
Because of the length involved, two
tug or towboats are required to
control the pipe as it is being taken to
location. Once on site, the buoyancy
modules are (carefully) removed or flooded, and the pipeline settles to the sea floor.

The *mid-depth* tow requires fewer buoyancy modules. Both the depth of submergence and
the shape of the pipeline are controlled by the forward speed of the tow. In this case, the pipeline
settles to the bottom on its own when the forward progression ceases.

The third case, *off-bottom tow*, involves both buoyancy modules and added weight in the form
of chains. The chains cause the pipeline to sink to near the bottom, but as the chain links pile up
on the sea floor, their weight reduces and the buoyancy holds the pipe at a given distance above
bottom. Again, once on location, the buoyancy is removed, and the line settles to the sea floor.

The last variation is *bottom tow*. In this case, the pipeline is allowed, or caused, to sink to the
bottom and then towed along the sea floor. This alternative is used primarily for soft and flat
bottoms, and in shallow water.

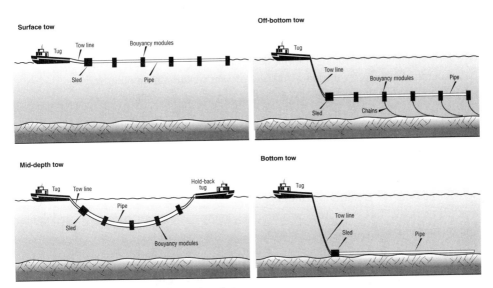

Fig. 10–10 Variations on Pipeline Tow-in Installation

In all of these approaches, the ends of the pipelines are configured with termination structures, often called sleds, so that one end can be connected to subsea manifolds and the other to the host platform riser system, using the previously discussed *jumper* method.

All of these methods have been used and are a way to do much work onshore and to limit exposure to offshore weather conditions. However long tow distances also have their own weather, shipping, and other risks.

Bottom Conditions

To a pipeline engineer, the ideal subsea terrain has a continuous flat seabed of soft clay or mud. Many deepwater areas have this type of seabed, but not all. The features of the ocean floor vary as much as the onshore, with gullies, outcrops, ravines, hills, and escarpments. The precursor to any pipeline or flowline installation is a survey by depth finding sonar and even ROV to establish the safest and most economical route. Even small undulations of the seabed cause worry. As the pipeline passes over an extended dip in the surface, it has an *unsupported span*, where the pipeline does not rest firmly or perhaps even touch the bottom. Bottom currents can lead to serious vibration in long unsupported spans.

Strong currents can find the gaps prime targets for scouring even longer expanses, making the pipeline more vulnerable. ROVs search for these unsupported spans. Additional weights or screw-type anchors can be added to push the line down or at least hold it steady. Alternatively, the spans can be outfitted with vortex shedding strakes to prevent vibration from occurring.

Risers

To and from the production platform, production and export risers take six different forms: attached risers, pull tube risers, steel catenary risers, top-tensioned risers, riser towers, and flexible riser configurations.

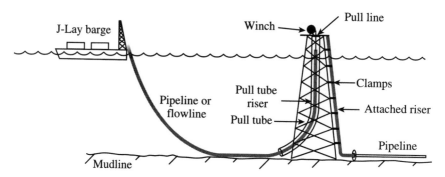

Fig. 10–11 Attached and Pull Tube Risers

Attached risers. Fixed platforms, compliant towers, and concrete gravity structures often have the risers clamped to the outside structure, as the right side of Figure 10–11 shows. The riser is prebuilt in sections. The sea bottom end of the riser is attached to the nearby inbound flowline or outbound pipeline, then assembled piece-by-piece and inserted into pre placed clamps on the structure. ROVs with electro-hydraulic tools make the connections or divers do the work.

Pull tube risers. In this installation, shown on the left side of Figure 10–11, a flowline or a pipeline becomes the riser as it is literally dragged up through a tube in the center of the structure. The pull tube, a few inches wider than the flowline or pipeline, is pre-installed in the structure. A flowline or pipeline laid on the seabed floor is attached to a wire rope threaded through the pull tube. The wire rope is winched up the pull tube to the topsides, at which point the flowline or pipeline inside the tube becomes the riser. This procedure works best, of course, when the pipe is pulled directly from the lay barge and fed into the pull tube.

Steel catenary risers. Some of the world's most graceful bridges, including the Golden Gate in San Francisco and the Verrazano Narrows in New York, exhibit a classic geometric shape. The cables that are strung from the end towers and support the bridge deck form a *catenary*, the mathematical formula shown in Figure 10–12. Another imposing and classic catenary, this

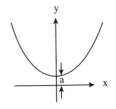

Catenary: the curve formed by a flexible cord or chain of uniform density, laying freely between two points at the same level.

The shape of the cord follows the formula in Cartesian coordinates,

$$y = \frac{a}{2}(e^{x/a} + e^{-x/a})$$

Fig. 10–12 Definition of a Catenary

time inverted, is the Gateway Arch in St Louis, Missouri. Designed by Eero Saarinen, it towers 630 ft above the Mississippi River. The steel catenary riser shown schematically in Figure 10–1 has a similar, albeit abbreviated, shape from the seabed to the production platform. As with the bridge cables, the more tension on the riser, the less sag in the system, and in the case of risers, the further out is the touchdown point.

Steel catenary risers are an elegant solution to the riser challenge of connecting to floating production platforms in deepwater. These risers can be installed by laying a predetermined length of pipe, making the connections easier. They tolerate a certain amount of movement of the production platform, making them useful for TLPs, FPSs, FPSOs, and spars, as well as fixed platforms, compliant towers, and gravity structures. However, excessive motion can cause metal fatigue at or near the touchdown point and at or near the top support.

Top-tensioned risers. For some TLP and spar applications, top-tensioned risers provide a good solution. In this application, the flowline or pipeline terminates at a junction point beneath the structure. A straight riser is run down from the platform and either terminates in the same junction structure, or in its own junction structure, in which case a short pipe section installed with the aid of an ROV connects the riser and flowline. (*See* Fig. 10–13.)

Since the riser is fixed to the sea floor, and the TLP or spar is free to move laterally under the influence of winds, waves and currents, there is vertical displacement between the top of the riser and the point where it connects to the TLP or spar. To accommodate this motion within allowable stresses in the riser, a hardware device known as a motion compensator is integrated into the top-tensioning system. This piece of hardware is made up of hydraulic cylinders and nitrogen-charged vessels known as accumulator bottles. (*See* Fig. 10–14.) An almost constant tension is maintained on the riser by the expansion-contraction action of the hydraulic fluid as it moves between the cylinders and the accumulator bottles. For spars, a simpler approach is often employed, using buoyancy cans around the outside diameter of the riser.

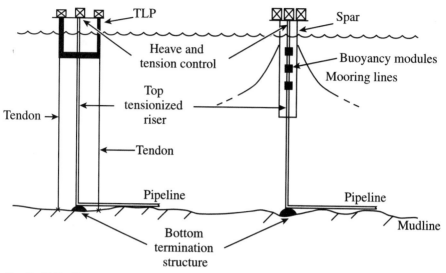

Fig. 10–13 Top-Tensioned Risers

The Christmas tree on top of the riser is, in either case, connected to the facilities manifold using flexible pipe, as seen in Figure 10–15, going from the motion compensator to the topsides facilities.

Riser towers. At the *Girassol Luanda* project offshore Angola, TotalFinaElf used the first riser towers. In this application, three steel column towers, each 4200 ft tall, are anchored to the sea floor in 4430 ft of water. The towers each have four production risers, four gas lift risers, two water-injection risers and two service line risers. The tower is topped with a buoyancy tank, 130 ft long and from 15 to 26 ft in diameter. The upward force generated by the buoyancy tank holds the tower vertically stable.

Flowlines connect from the subsea wells to the tower base, and flexible lines from the nearby FPSO connect to the various risers at the tower top.

Flexible riser configurations. Petrobras pioneered the use of flexible pipe as risers starting in 1978. Because of the bending characteristics, flexibles are most appropriate for floating systems that experience significant vertical and/or horizontal movement.

Flexible pipe, because of its beneficial bending characteristic, is often used to make the connections from the production equipment on floating systems to the production and export risers. Because of these same bending characteristics, these flexibles are also finding more and more use as the primary risers.

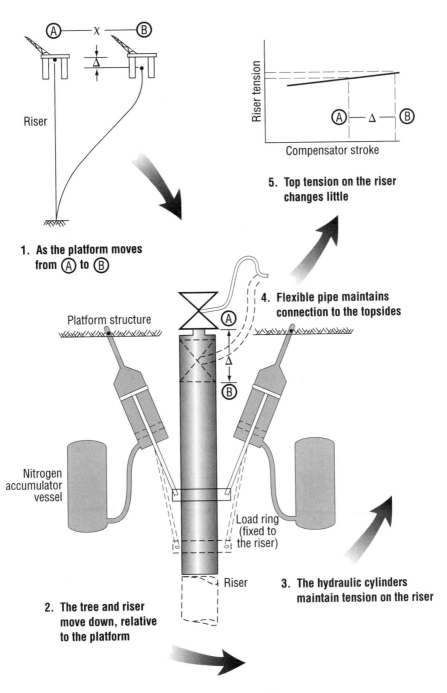

1. As the platform moves from Ⓐ to Ⓑ

Riser

5. Top tension on the riser changes little

4. Flexible pipe maintains connection to the topsides

Platform structure

Nitrogen accumulator vessel

Load ring (fixed to the riser)

Riser

3. The hydraulic cylinders maintain tension on the riser

2. The tree and riser move down, relative to the platform

10–14 Motion Compensation Mechanism for Top Tension Risers

Fig. 10–15 Riser Connections at the *Auger* TLP (Note the flexible hoses between the motion compensators and the topsides facilities; courtesy FMC Technologies)

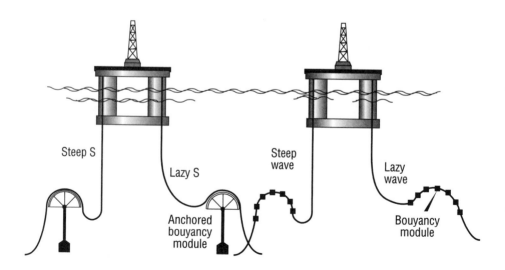

Fig. 10–16 Some Flexible Riser Configurations

Besides the usual, free-hanging catenary riser, several other configurations are used. Figure 10–16 shows four: lazy S, steep S, steep wave, and lazy wave. The specific selection depends on the expected motions of the production vessel and available installation equipment. All configurations involve providing buoyancy modules or cans at selected locations on the flexible pipe to force a given shape, and they all have the net effect of reducing the stress and wear and tear on the pipe at both the touchdown point and at the platform connection.

Selecting a risers system

For the various development systems, some risers predominate. For fixed platforms and compliant towers, the attached and the pull-tube risers are most common. For concrete gravity structures, attached- and pull-tube risers make most sense. For floating systems like FPS, TLP, FPSO, and spar, the attached riser and pull tube don't work because they cannot accommodate the lateral excursions. For the floater, the steel catenary riser and flexibles work for almost all occasions. The top-tensioned riser works for the TLP and spar only. The riser tower method, which also has many variations, can be used with all floating systems, particularly for ultra deepwater.

Pipeline System Operations

Once all the lines are hooked up, the valves are opened up, and oil and gas is moving to market, then *flow assurance* (keeping the fluids moving through the lines) moves towards the top of the watch list on the platform. Operating crews monitor pressures and flows, watch for potential plugging, and inject inhibiting chemicals when necessary. From time to time, operators sweep out flowlines and pipelines by sending a *pig*, a specially designed sphere or short cylinder, down the line to sweep out the line. The pressure of the oil or gas drives the pig as it scrapes layers of paraffin off the pipeline walls or pushes ahead of it hydrates, water, and accumulated sand or other trash.

Pig launch and recovery facilities that are aided by ROVs handle the start and finish. Pigs can be simple constructs of polyurethane or *smart pigs* that not only clean but also have sensors to measure wall thickness, monitor corrosion, and check for faults and leaks.

PIGS?

Where did the name come from? An expert in the field, Andrew Marwood of Pipeline Engineering Company, offers two possibilities:

- PIG, a Pipeline Injected Gadget
- In the early days of pipelining, leather balls were used to swab pipelines. As they moved down the line they squealed—like a pig.

Marwood suggests that both undocumented possibilities may be more a myth of mirth than a matter of fact.

11

Technology and the Third Wave

And in the lowest deep a lower deep.

From *Paradise Lost*

John Milton (1608–1674)

It took the oil and gas industry 50 years of inching up the offshore learning curve before Kerr-McGee placed the historic Creole Platform in 30 ft of Gulf of Mexico waters in 1947. In the next half century, a remarkable technology story unfolded as industry soared nearly to the top of the curve, to the point where Shell began production in 1989 from a platform in 1354 ft of water.

Even before the curve started to flatten out, offshore pioneers leaped ahead. In the next decade, they routinely produced oil and gas from wells in 2000–3000 ft of water in the Gulf of Mexico and the Campos Basin. Soon they drilled wells in 5000–10,000 ft. The oil and gas industry had moved to the steep part of the deepwater learning curve, the third wave.

The previous 10 chapters have documented the most recent accomplishments in the years 1970 to 2000, all vital to success in the deepwater. Some belong to the mature offshore wave, some even the fully ripened first wave, the onshore.

- Horizontal wells
- Bright spots
- High capacity computing
- 4D seismic
- FPSs
- CTs
- Subsea wells
- Tension leg platforms
- Spars

- FPSOs
- J-lay pipeline vessels
- S-lay pipeline vessels
- Large capacity lift barges, supply boats, and service vessels
- Taut anchoring systems
- Catenary risers
- Frac Pack completions
- Suction piles
- High rate/high reserve wells

Even as this book reaches publication, technologies continue to emerge:

- Shear seismic — Shear waves, oscillations of particles perpendicular to the sound source (and to the P-waves generally used in offshore seismic), do not transmit through water. Capturing S-waves allows more detailed description of subsurface properties.
- Riserless drilling — The weight of the drilling mud in the riser annulus increases the frequency that casing has to be run. Pumping the returning mud separately from the sea floor to the drilling platform eliminates part of the weight.
- Composite materials — Using composite materials—polymers and reinforced resins—reduces weight in platforms, casing, risers, anchor lines, and other applications.
- Smart wells — Wells that have analytical apparatus in the wellbore and at the wet tree are controlled from remote locations.
- Expandable, monobore tubulars — An elastic steel casing that can be expanded after it is inserted in the wellbore permits a well of uniform diameter, top to bottom. Elimination of the need for wider casing in the upper sections of the well reduces materials and cost.
- Dual activity drilling rigs — Drilling and setting wellheads simultaneously improves the rig efficiency.
- Subsea pumping — Pumps on the sea floor, connected to a power source on a remote platform, extend the range of subsea completion options and make smaller reservoirs economical.
- Subsea separation — Placing oil/gas/water separators on the sea floor increases the options for where and how those fluids get handled at the surface.
- Surface blowout preventer — Having the BOP on the platform makes maintenance (unjamming rams, replacing seals, etc.) more efficient but requires high pressure risers and subsea failsafe devices.
- Autonomous underwater vehicles (AUVs) — Not having an umbilical, the AUV is unencumbered in depth and maneuvering; only battery capacity limits the range. An AUV gets recharged without returning to the surface by connecting to a seabed station linked to a platform that provides electrical power.

And as operating companies deliver those technologies to deepwater sites, research organizations of a few oil and gas companies and scores of service companies aggressively pursue the means to bring hydrocarbons from ever more inhospitable locations and conditions in the deep- and ultra deepwater:

- Direct detection of hydrocarbons below salt — dealing with the distortions as seismic waves pass through salt.
- Real time seismic acquisition, processing, and interpretation — cutting the cycle time from acquisition to decision.
- Light weight, zero discharge rigs and processing facilities — addressing with increasing demands of regulatory agencies.
- Totally composite structures — stronger, lighter, less expensive facilities by replacing steel.
- Exploiting gas hydrates — recovering the trillions of cubic feet of methane locked up in deep methane hydrate deposits.
- Tapping small accumulations — the combination of multiple technologies allowing efficient access and evacuation.
- Floating gas liquefaction plants and gas-to-liquids technology — exploiting stranded gas accumulations around the world using an economic means of transport to market, either
  - by liquefying the methane and moving it as LNG, or
  - by chemically converting it to a diesel range oil.

Prospects

Despite the spectacular advances, deepwater is an immature frontier. As this book goes to print, industry has found about 60 billion barrels of hydrocarbons in waters deeper than 1500 ft. More than half was found since 1995, with "Golden Triangle," the Gulf of Mexico, the Campos Basin, and West Africa accounting for most of the volume. A score of other prospective areas around the world (Fig. 11–1) remain mostly unexplored, especially

Fig. 11–1
Deepwater Areas with Hydrocarbons Potential (shown in the shaded areas)

- The Faroe Islands, the West of Shetlands, and mid-Norway
- Morocco, Egypt, the Adriatic, and eastern Mediterranean
- South Caspian
- Trinidad to the Falklands
- Nova Scotia
- Tanzania and Mozambique
- Northwest and West South Africa
- Northwest Australia, South Australia, and New Zealand
- West and East India
- Borneo and the Philippine Sea
- Sakhalin

Some of these areas are down-dip (further out, in deeper water and deeper sediment) from existing producing areas, both onshore and shallow offshore. Others are brand new prospects. Many should have the same, wonderfully productive turbidite reservoirs that have made existing deepwater ventures so successful.

Ultimately

M. King Hubbert, a brilliant but irascible geophysicist, forecasted in 1956 that U.S. oil production would peak in 1970, plus or minus a couple years, following the trajectory of a bell-shaped curve. At the time, of course, he was looking at only the upside of that curve. Incredibly enough, U.S. production began to decline in 1971. Hubbert's curve, as it came to be known, amazed and excited petroleum economists around the world.

In ensuing years, Hubbert and his disciples used his notions about finite resources to predict that *world* oil production would peak by 2000. The Hydrocarbon Age would be over. As that date approached, appraisals of the ultimately recoverable oil, the area under Hubbert's curve, increased and Hubbert subscribers postponed their gloomy prognoses further into the future. Some still predict peak world oil production in the first decade of the 21st century.

The technology learning curve, the wave, used so enthusiastically in this book, is of course, the integral of a bell-shaped curve. That lends some support to Hubbert's theory, but only to the extent that oilmen embark on no new waves, no new learning curves. In an interminable public debate with Hubbert subscribers, Professor Morris Adelman and his colleagues at the M.I.T Center for Energy and Environmental Policy argue that what limits the supply of oil and gas is not any finite resource theory but only the ingenuity of the people searching for and producing them. And, that, they maintain, is limitless. Perhaps the deepwater wave is just one of a series that will keep us in the Hydrocarbon Age indefinitely.

Index

A

Absorption process, 131

Abyssal zone, 49

Acid gases, 126

Acoustic equipment, 20–22

Acquisition (seismic data), 41–43

Acquisition (lease), 37–39

Acronyms, xiv

Adriatic Sea, 158

Amplitude versus offset (seismic data), 45

Anchor/anchoring, 103, 156

Annulus (well), 59

Appraisal (well), 47–48

Associated gas, 75

Attached risers, 148

Australia, 158

Authorization for expenditure (AFE), 55, 61, 63

Autonomous underwater vehicles, 156

Auxiliary systems, 136–138:
power, 136–137;
pumps, 137;
cranes, 137;
flares, 137–138;
export connections, 138

B

Basic sediment and water content, 126, 128

Bid/bidding, 38, 57

Blowout preventers (BOP), 59–61:
rams, 60;
stack, 60

Boilers, 138

Bootstrapping, 9–11

Borneo, 158

Bottom conditions, 147–148

Bottom tow, 146–147

Brazil, 31–33

Bright spots, 21, 155

Buoyancy (underwater pipeline), 141–142

C

Caissons, 100–101

Campos Basin (Brazil), 31–33, 157

Caspian Sea, 158

Catenary risers, 149, 153, 156

Cell spar, 99

Choke line, 60

Christmas trees, 64, 150

Coalescers, 129

Coiled tubing, 155

Commingled fluids, 140

Completion (well), 28, 61–66, 155–156:
prognosis, 61–63;
mechanicals, 63–66;
completion fluid, 63–64, 66;
downhole, 64–65;
completion fluid removal, 66

Completion fluid, 63–64, 66:
removal, 66

Completion prognosis, 61–63

Compliant towers, 70–71, 79–81, 83–85:
construction, 80–81;
installing, 83–85

Composite materials, 156–157

Concrete platforms, 3–4, 70–71, 78–81, 83, 85:
gravity platforms, 81;
transportation, 83;
installing, 85

Conductor pipe, 77

Confusion factors (1980s), 27–30

Construction materials, 3–4, 81, 156–157

Construction/installation (FPS), 100–104:
ship-shape hulls, 100;
caissons, 100–101;
pontoons, 100–101;
spars, 101;
transportation, 101–103;
installation, 103–104;
setting the deck, 104

Continental shelf, 49–51

Control pods, 113

Control systems, 109, 113–114

Conventional platforms, 77–78, 83–85:
    installing, 83–85

Cranes (topsides), 18–19, 137, 156

Creole field (LA), 5–6

Crew capsules, 135

Crew size/complement, 133

Crude content, 126

Current velocity, 67–68

CUSS concept, 11–13

D

Data acquisition (seismic), 41–43

Data display (seismic), 41, 43, 46

Data interpretation (seismic), 41, 46

Data preparation (seismic), 43

Data processing (seismic), 41–45:
    preparation, 43;
    stacking, 43;
    migration, 44–45;
    prestack depth migration, 45;
    direct hydrocarbon indicators, 45;
    amplitude versus offset, 45;
    four-dimensional seismic, 45

Dead oil, 127–128

Deck, 77–78, 85–87, 104:
    setting the deck, 85–87, 104

Deepwater discoveries, 28–29, 31–34

Deepwater exploration, 28–29, 31–51:
    discoveries, 28–29, 31–34;
    developing a play, 36–37;
    acquisition, 37–39;
    identifying the prospect, 39–46;
    seismic data, 40–46;
    wildcat drilling, 47–48;
    deepwater plays, 48–49;
    geology, 49–51

Deepwater plays, 48–49

Deepwater problems, 66–68:
    loop current/eddy current, 66–68;
    shallow depth hazards, 68;
    shallow water flows, 68

Deepwater prospects, 48–51, 157–158:
    plays, 48–49;
    Gulf of Mexico, 49–51

Deepwater technology, 155–158

Deepwater well costs, 51

Definitions, xiv, 37

Dehydration (gas), 131–132

DeLong design, 9–11

Design storm, 91

Dessicant process, 131

Developing a play, 36–37

Development systems, 69–76:
    fixed platforms, 70–71;
    floating systems, 70–72;
    subsea systems, 70, 72;
    choices, 70–76;
    selection process, 72–76

Direct hydrocarbon indicators, 45

Display (seismic data), 41, 43, 46

Divers/remotely operated vehicles, 16–19

Diving technology, 16–19

Double hull, 95–96

Downhole completion, 64–65

Drake's well (PA), 25–26

Drilling (well), 1–16, 28, 53–61, 96, 156:
    history, 1–5;
    mobile rigs, 4–5;
    platforms, 6–16;
    drilling tenders, 7–8;
    drilling ships, 11–16, 28, 55–56, 58, 96;
    technology, 28;
    drilling process, 53–54;
    well plan, 54–55;
    prognosis, 54–55;
    rig selection, 55–57, 156;
    drilling operations, 57–61;
    drilling mud, 58–59;

drillstem test, 61;
FDPSO, 96

Drilling mud, 58–59

Drilling operations, 57–61:
    drilling mud, 58–59;
    blowout preventers, 59–61

Drilling platforms, 6–16

Drilling process, 53–54

Drilling prognosis, 54–55

Drilling rig selection, 55–57

Drilling ships, 11–16, 28, 55–56, 58, 96:
    dynamic positioning, 28

Drilling technology, 28

Drilling tenders, 7–8

Drillstem test, 61

Dry gas, 131

Dry oil, 128

Dry tow, 102

Dry trees, 64, 99

Dual activity drilling rigs, 156

Dynamically-positioned drillships, 28

E

Eddy current, 66–68

Effluent control, 157

Egypt, 158

Electric generators, 136

Electricity, 136–138

Elephant hunt, 32

Elwood field (CA), 2

Escape capsules, 135

Evaluating the well, 61

Expandable well, 55

Expandable/monobore tubulars, 156

Exploratory drilling, 47–48:
    appraisal, 47–48

Export connections (topsides), 138

Export riser, 140

F

Falkland Islands, 158

Faroe Islands, 74, 158

FDPSO, 96

Fixed platforms, 6–16, 22–24, 70–71, 77–87:
sea floor, 71;
platform jacket, 71;
compliant towers, 71, 79–81;
gravity platforms, 71;
fixed structures, 77–87;
conventional platforms, 77–78;
concrete platforms, 78–81;
transportation, 81–83;
installing, 83–87

Flares (topsides), 137–138

Flash (natural gas), 127

Flex trend, 50

Flexible pipe, 141, 150, 152–153:
risers, 150, 152–153

Flexible riser configurations, 150, 152–153

Floating drilling, production, storage, and offloading system. SEE FDPSO.

Floating pile driver, 3

Floating production systems (FPS), 3, 11–16, 31–32, 70–72, 74, 89–106, 155–156:
FPSOs, 32, 70, 72, 74, 93–96, 156;
tension leg platforms, 71, 91–93;
spar platforms, 72, 98–100;
options, 89;
hull, 90;
topsides, 90;
mooring, 90, 97, 104–106;
risers, 91, 106;
tanker conversion, 94;
tanker offloading, 96;
FDPSO, 96;
application, 97;

construction and installation, 100–104

Floating production, storage, and offloading systems. SEE FPSO.

Floating storage and offloading (FSO), 96, 100

Float-over method, 104

Flow assurance, 114–115, 153

Flowlines, 109, 111–112, 115, 140, 142, 150

Flying leads, 112–113

Formation fracture gradient, 59

Foundation piles, 104

Foundation sleeves/slots, 84

Four-dimensional seismic, 45, 155

FPSO, 32, 70, 72, 74, 93–96, 156:
tanker conversion, 94;
tanker offloading, 96

Frac pack completions, 156

Free span, 147

Fuel gas, 136, 138

G

Gambler's ruin, 48

Gas condensate, 130

Gas content, 126

Gas flaring, 137–138

Gas hydrates, 114, 129–130, 157

Gas liquefaction plants, 157

Gas metering, 132–133: meter types, 132–133

Gas treatment, 129–133:
heating, 129–130;
separation, 130;
stabilization, 130–131;
dehydration, 131–132;
metering, 132–133

Gas updip, 51

Gas-to-liquids technology, 157

Gathering lines, 111–112

Geochemistry, 37

Geologic forces, 36

Geologic problems, 68

Geologic trap, 36

Geology/geophysics, 18, 20–22, 37, 49–51:
Gulf of Mexico, 49–51

Geophones, 41–43

Geophysics, 20–22, 37, 40–46, 156:
seismic exploration, 20–22;
seismic data, 40–46;
shear seismic, 156

Giliasso design, 4

Global positioning system (GPS), 17, 117

Glycol dehydration system, 131

Government policy, 72

Gravel pack operations, 65–66

Gravity platforms, 71, 78–79, 81, 85:
installing, 85

Gulf of Mexico, 1–11, 27–30, 48–51, 66–68, 72–74, 157:
geology, 49–51

H

Header, 111

Heating (gas), 129–130

Heave compensation devices, 115

Heavy lifting, 18–19

Helicopters, 134

High capacity computing, 155

History (petroleum industry), 1–26:
early offshore producing, 1–5;
offshore structure concepts, 6–16;
divers and remote operated vehicles, 16–19;
geology and geophysics, 18, 20–22;
prefabricated structures, 23–24;
offshore learning curve, 25–26

Horizontal wells, 155

Hubbert prediction, 158

Hull (platform), 89

Hydrate formation, 129–130

Hydril annular preventer, 60

Hydrocarbon detection, 157

Hydrocarbon potential, 157

Hydrocyclones, 129

Hydrodynamics, 66–68, 141–142, 147–148

Hydrophones, 41–43

I

Identifying prospect, 39–46:
seismic data, 40–46

India, 158

Installation coordinator, 117

Installing platforms, 83–87, 91–92, 103–104, 115–119:
conventional platforms, 83–85;
compliant towers, 83–85;
concrete platforms, 85;
gravity platforms, 85;
setting the deck, 85–87;
setting the riser, 87;
tension leg platforms, 91–92;
subsea systems, 115–119

Integrated umbilical, 112

Interpretation (seismic data), 41, 46

Investment preferences, 72

J–K

Jackdown concept, 10–11

Jackups, 7, 9–11

J-lay configuration (pipeline), 142, 144, 156

Jumpers, 111–112, 118, 139, 142, 147

L

Lake Caddo fields (TX), 2–3

Lake Maracaibo fields (Venezuela), 3–4

Lake Pelto field (LA), 4–5

Landing ship tank, 7

Launch barge, 81

Lay barge (pipeline), 141–147, 156

Learning curve, 25–26

Lease acquisition, 37–39

Lease automatic custody transfer, 132

Lift power, 18–19

Lifting (platform), 18–19, 81, 156

Lithological trap, 36

Lithology (definition), 37

Live oil, 127

Load out, 81

Logistics (platform installation), 77–87

Loop current, 66–68

Low temperature, 140–141

Lowering, 85

M

Maneuvering (ROV), 121–122

Manifolds, 109–111, 117–118

Manipulators, 120

Marginal wells, 157

McFadden Beach project (TX), 5

Measure while drilling, 58

Mechanicals (well completion), 63–66

Mediterranean Sea, 158

Meter provers, 132

Metering (gas), 132–133:
meter types, 132–133

Meters, 132–133:
metering, 132–133;

meter provers, 132;
positive displacement meter, 132;
turbine meter, 132;
ultrasonic meter, 132;
orifice meter, 132–133

Mexico, 39

Mid-depth tow, 146–147

Migration (hydrocarbon), 36

Migration (seismic data), 44–45

Minerals Management Service, 38

Mobile offshore drilling, 4–5

Modern era, 27–34:
confusion factors, 27–30;
rig activity, 27–29;
new directions, 31–34

Monobore tubulars, 156

Monocolumn TLP, 92–93

Mooring hawser, 96

Mooring, 15–16, 70–71, 90–91, 93, 95–97, 100, 104–106:
hawser, 96;
spread configuration, 104–106

Morgan City project (LA), 5

Morocco, 158

Mozambique, 158

Mud mats, 84

Mud removal, 63

Mud weight, 59

Multiphase fluid, 125–128

N

Natural gas liquids, 131

New directions, 31–34

New Zealand, 158

North Sea, 72–74, 78

Norway, 158

Nova Scotia, 158

O

Off-bottom tow, 146–147

Offloading (oil), 72, 96

Offloading hose, 96

Offshore completion, 28, 61–66,
155–156:
prognosis, 61–63;
mechanicals, 63–66;
completion fluid, 63–64, 66;
downhole, 64–65;
completion fluid removal, 66

Offshore drilling, 1–16, 28, 53–61,
96, 156:
history, 1–5;
mobile rigs, 4–5;
platforms, 6–16;
drilling tenders, 7–8;
drilling ships, 11–16, 28, 55–56,
58, 96;
technology, 28;
drilling process, 53–54;
well plan, 54–55;
prognosis, 54–55;
rig selection, 55–57, 156;
drilling operations, 57–61;
drilling mud, 58–59;
drillstem test, 61;
FDPSO, 96

Offshore producing (early era), 1–5

Offshore storage, 32, 70, 72, 74,
93–96, 100, 156:
FPSOs, 32, 70, 72, 74, 93–96,
156;
tanker conversion, 94;
tanker offloading, 96;
FSO's, 96, 100

Offshore structure concepts, 6–16:
submersibles, 8–9;
bootstrapping, 9–11;
jackups, 9–11;
floaters, 11–16;
positioning, 14, 16

Oil skimmers, 128–129

Oil transportation, 72, 74, 96

Oil treatment, 127–129:
oil-gas separation, 127–128;
oil-water separation, 128–129;

treaters, 128;
skimmers, 128–129

Oil-gas separation, 127–128

Oil-water separation, 128–129

OPEC, 27–28

Operators (ROV), 123

Orifice meter, 132–133

P

Paleontology, 47

Paraffin inhibitors, 115

Passive margin, 49

Permeability (rock), 36–37

Personnel and quarters, 133–134:
crew size/complement, 133;
shifts, 133;
quarters, 133–134;
transportation, 134

Petroleum industry history, 1–26:
early offshore producing, 1–5;
offshore structure concepts,
6–16;
divers and remotely operated
vehicles, 16–19;
geology and geophysics, 18,
20–22;
prefabricated structures, 23–24;
offshore learning curve, 25–26

Petrophysics, 47

Philippine Sea, 158

Pigs/pigging, 153

Piling/pile driving, 3, 77, 84–85,
104:
floating pile driver, 3

Pipe rams, 60

Pipe-in-pipe flowline, 114,
140–141

Pipeline construction, 142–147,
156:
pipeline laying vessels, 142–147,
156;
S-lay configuration, 142–143;
J-lay configuration, 142, 144;

reel barge, 142, 145–146;
tow-in method, 142, 146–147

Pipeline design, 140–141

Pipeline infrastructure, 74–75

Pipeline laying vessels, 142–147,
156:
S-lay configuration, 142–143;
J-lay configuration, 144;
reel barge, 145–146;
tow-in method, 146

Pipeline pig/pigging, 153

Pipeline system operations, 153

Pipelines, 74–75, 111–112, 114,
139–153, 156:
infrastructure, 74–75;
pipe-in-pipe flowline, 114,
140–141;
pipeline design, 140–141;
buoyancy, 141–142;
construction, 142–147, 156;
pipelaying vessels, 142–147,
156;
bottom conditions, 147–148;
risers, 148–153;
pigging, 153;
pipeline system operations, 153

Plate coalescers, 129

Plate tectonics, 37

Platform deck, 77–78, 85–87, 104:
setting the deck, 85–87, 104

Platform jackets, 7, 71, 77–78

Platform rigs, 57

Play, 36–37:
development, 36–37

Plugged lines, 114

Pontoons, 100–101

Porosity (rock), 36–37

Positioning (location), 12, 14–17,
83, 117:
global positioning systems, 17,
117

Positive displacement meter, 132

Power systems (topsides), 136–137

Prefabricated structures, 6, 23–24

Prestack depth migration (seismic data), 45

Process (completion), 61–63

Process (drilling), 53–54

Process controls, 134–135

Production peak, 158

Production platforms, 3, 6–16, 22–24, 31–32, 70–72, 77–87, 89–106, 155:
floating platforms, 3, 11–16, 31–32, 70–72, 89–106, 155;
fixed platforms, 6–16, 22-24, 70–71, 77–87

Production risers, 87–88, 91, 106, 140, 148–153, 156:
setting, 87;
attached, 148;
pull tube, 148;
steel catenary, 149, 153, 156;
top-tensioned, 149–152;
riser towers, 150;
flexible riser configurations, 150, 152–153;
selection, 153

Prognosis, 54–55, 61–63:
drilling, 54–55;
completion, 61–63

Project development, 69–76:
development system choices, 70–76

Prospect appraisal, 47–48

Prospect evaluation, 35–51:
developing a play, 36–37;
lease acquisition, 37–39;
identifying the prospect, 39–46;
seismic data, 40–56;
wildcat drilling, 47–48;
appraisal, 47–48;
deepwater plays, 48–49;
geology, 49–51

Prospect identification, 39–46:
seismic data, 40–46

Pull tube risers, 148

Pumps (topsides), 137

Q

Quarters (personnel), 133–134:
crew size/complement, 133;
shifts, 133;
quarters, 133–134;
transportation, 134

R

Rams (BOP), 60

Rathbone's well (WV), 26

Real time seismic, 157

Reel barge, 142, 145–146

Refrigeration, 131–132

Regeneration (glycol), 131

Reliability (subsea system), 119

Remote control, 109, 113–114

Remote operated vehicles (ROV), 16–19, 92, 106, 110, 113–114, 117–123:
maneuvering, 121–122;
umbilicals, 122–123;
tethering, 122–123;
operators, 123

Reservoir characteristic, 36

Reservoir proximity, 76

Revolving turret, 95

Rig activity, 27–29

Rig capability, 55–57, 156

Rig selection, 55–57

Riser towers, 150

Riserless drilling, 156

Risers, 87–88, 91, 106, 140, 148–153, 156:
setting, 87;
production, 140;
attached, 148;
pull tube, 148;
steel catenary, 149, 153, 156;
top-tensioned, 149–152;
riser towers, 150;
flexible riser configurations, 150, 152–153;
selection, 153

S

Safety features, 64, 133

Safety systems, 134–135

Sakhalin Island, 158

Salt deposits, 50–51

Saturation diving, 18

Sealing mechanism, 36

SeaStar TLP, 92–93

Seismic data, 20–22, 40–46, 156:
exploration, 20–22;
data acquisition, 41–43;
data processing, 41–45;
data display, 41, 43, 46;
data interpretation, 41, 46;
records, 42;
trace, 42–43;
shear seismic, 156

Seismic exploration, 20–22

Seismic records, 42

Seismic trace, 42–43

Selection process (completion), 72–76:
water depths, 73–74;
oil transportation, 74;
gas disposition, 75;
reservoir proximity, 76

Semisubmersibles, 13–14, 55–56, 58

Separation (gas), 130

Separation (oil), 128

Setting the deck, 85–87, 104

Setting the riser, 87

Sewage treatment, 138

Shallow depth hazards, 68

Shallow water flows, 68

Shatto's calculation, 15

Shear rams, 60

Shear seismic, 156

Shifts (personnel), 133

Ship Shoal area, 6–7

Ship-shape hulls, 100

Shut-in wells, 114–115

Shuttle tanker, 96

Sidetracking, 55

Skimmers, 128–129

S-lay configuration (pipeline), 142–143, 156

Sleds, 109–111

Smart pigs, 153

Smart wells, 156

Source rock, 36–37

South Africa, 158

Spar platforms, 70, 72, 98–101, 103–104, 155:
truss spar, 99;
cell spar, 99

Spider buoy, 95

Spindletop field (TX), 25

Spread configuration (mooring), 104–106

Spud the well, 57

Stabilization (gas), 130–131

Stabilizer (condensate), 130

Stack (BOP), 60

Stacking (seismic data), 43

Steel catenary risers, 149, 153

Steel platform, 70

Storage capacity, 95

Strakes, 98

Stranded gas, 75, 157

Streamers, 41

Structure transportation, 81–83

Submersibles, 4–5, 8–9

Subsea completion, 28, 61–66, 155–156:
prognosis, 61–63;
mechanicals, 63–66;
completion fluid, 63–64, 66;
downhole, 64–65;
completion fluid removal, 66

Subsea pipelines, 139–153:
pipeline design, 140–141;
buoyancy, 141–142;
laying pipe, 142–147;
bottom conditions, 147–148;

risers, 148–153;
pipeline system operations, 153

Subsea pumping, 156

Subsea riser basket, 92

Subsea separation, 156

Subsea systems, 107–123:
underwater wells, 109;
trees, 109–110;
manifolds, 110–111, 117–118;
sleds, 110–111;
underwater pipelines, 111–112;
flowlines, 111–112;
jumpers, 111–112, 118;
gathering lines, 111–112;
umbilicals, 112–113, 118–119, 122–123;
flying leads, 112–113;
control systems, 113–114;
flow assurance, 114–115;
system architecture, 115–119;
installation, 115–119;
reliability, 119;
remote operated vehicles, 119–123

Subsea trees, 109–110

Subsea vehicles, 16–19, 92, 106, 110, 113–114, 117–123

Subsea wells, 28, 61–66, 109–110, 155:
completion, 28, 61–66, 155–156;
trees, 109–110

Subsurface safety valve, 64

Suction piles, 104–105, 156

Summerland field (CA), 1–2

Surface BOP, 156

Surface casing, 59

Surface tow, 146–147

Swivel stack, 95

System architecture (subsea), 115–119

T

Tanker conversion, 93–94

Tanker offloading, 96

Tanzania, 158

Target depth, 61

Taut anchoring systems, 156

Technology, 155–158:
deepwater prospects, 157–158;
production peak, 158

Template platforms, 6–7

Tendons, 91, 93, 103

Tension leg platforms (TLP), 32–33, 70–71, 91–93, 155:
installation, 91–92;
monocolumn, 92–93

Teredo worm, 3–4

Terminology, xiv

Tethering, 70–71, 91, 93, 122–123:
ROVs, 122–123

Three-dimensional seismic, 22, 28, 46:
data, 46

Topsides, 85–87, 90, 125–138:
oil treatment, 127–128;
water treatment, 128–129;
gas treatment, 129–133;
personnel and quarters, 133–134;
safety systems, 134–135;
auxiliary systems, 136–138

Top-tensioned risers, 149–152

Tow-in method, 142, 146–147

Towing, 82–83, 101–103, 142, 146–147:
wet tow, 102;
dry tow, 102;
tow-in method, 142, 146–147;
surface tow, 146–147;
mid-depth tow, 146–147;
off-bottom tow, 146–147;
bottom tow, 146–147

Transportation (oil), 72, 74, 96

Transportation (personnel), 134

Transportation (structure), 81–83, 101–103

Trap (hydrocarbon), 36

Treaters (oil), 127–129:
oil-gas separation, 127–128;
oil-water separation, 128–129;
design, 128;
skimmers, 128–129
Tree (well), 61–62, 64, 109–110:
Christmas trees, 64;
wet trees, 64, 109–110;
dry trees, 64;
subsea, 109–110
Trinidad, 158
Truss spar, 99
Turbidite sands, 32–33
Turbine meter, 132
Turrets, 94–95

U

Ultrasonic meter, 132
Umbilical termination structure, 113
Umbilicals, 109, 112, 113, 118–119, 122–123:
integrated, 112;
termination structure, 113
Underwater pipelines, 74–75, 111–112, 114, 139–153, 156:
infrastructure, 74–75;
pipe-in-pipe flowline, 114, 140–141;
pipeline design, 140–141;
buoyancy, 141–142;
construction, 142–147, 156;
pipelaying vessels, 142–147, 156;
bottom conditions, 147–148;
risers, 148–153;
pigging, 153;
pipeline system operations, 153
Underwater wells, 109–110, 155:
wellhead, 109–110
Unsupported span, 147

V

Vapor pressure, 130
Visualization (data), 46
Voxel technology, 46

W–Z

Water currents, 66–68, 147–148:
eddy currents, 66–68;
current velocity, 67–68
Water depths, 72–74, 102
Water downdip, 51
Water supply, 138
Water treatment, 128–129
Weathervane, 94
Well completion fluid, 63–64, 66:
removal, 66
Well completion, 28, 61–66, 155–156:
prognosis, 61–63;
mechanicals, 63–66;
completion fluid, 63–64, 66;
downhole, 64–65;
completion fluid removal, 66
Well costs, 51
Well plan, 54–55
Well testing, 47
West Africa, 74, 157
West of the Shetlands, 74, 158
Wet tow, 102
Wet trees, 64, 109–110
Wildcat drilling, 39, 47–48:
appraisal, 47–48
Wildcat well, 39